Risk Management Framework for Fourth Industrial Revolution Technologies

This book focuses on major challenges posed by the Fourth Industrial Revolution (4IR), particularly the associated risks. By recognizing and addressing these risks, it bridges the gap between technological advancements and effective risk management. It further facilitates a swift adoption of technology and equips readers with the knowledge to be cautious during its implementation. Divided into three parts, it covers an overview of 4IR and explores the risks and risk management techniques and comprehensive risk management framework specifically tailored for the 4IR.

Features:

- Establishes a risk management framework for Industry 4.0 technologies.
- Provides a 'one stop shop' of different technologies emerging in the Fourth Industrial Revolution.
- Follows a consistent structure for each key Industry 4.0 technology in separate chapters.
- Details required risk management skills for the technologies of the Fourth Industrial Revolution.
- Covers risk monitoring, control, and mitigation measures.

This book is aimed at graduate students, technology enthusiasts, and researchers in computer sciences, technology management, business management, and industrial engineering.

Risk Management Framework for Fourth Industrial Revolution Technologies

Omoseni Oyindamola Adepoju[1],
Nnamdi Ikechi Nwulu[2], and Love Opeyemi David[3]
[1]Department of Management and Accounting, School of Social Sciences and Management, Lead City University, Ibadan, Oyo state, Nigeria
[1,2,3]Centre for Cyber-Physical Food, Energy & Water Systems University of Johannesburg, South Africa

CRC Press is an imprint of the
Taylor & Francis Group, an informa business

Designed cover image: © Shutterstock Images

First edition published 2025
by CRC Press
2385 NW Executive Center Drive, Suite 320, Boca Raton FL 33431

and by CRC Press
4 Park Square, Milton Park, Abingdon, Oxon, OX14 4RN

CRC Press is an imprint of Taylor & Francis Group, LLC

ISBN: 9781032713779 (hbk)
ISBN: 9781032862682 (pbk)
ISBN: 9781003522102 (ebk)

DOI: 10.1201/9781003522102

Typeset in Times
by Apex CoVantage, LLC

Contents

PART III Risk Management Framework for Fourth Industrial Revolution Technologies

About the Authors

Professor Omoseni Oyindamola Adepoju is an Associate Professor at Lead City University Ibadan, Nigeria. Additionally, she serves as a Senior Research Associate within the Faculty of Engineering and Built Environment at the University of Johannesburg, South Africa. In 2020, she completed a post-doctoral research project focused on developing a rapid re-skilling framework for Construction 4.0 in Sub-Saharan Africa. This research was conducted at the Department of Construction Management and Quantity Surveying, University of Johannesburg, South Africa.

During her post-doctoral tenure, Professor Adepoju contributed significantly to the academic landscape, publishing multiple articles and a book on the topic of reskilling human resources for Construction 4.0, with implications for industry, academia, and government. Subsequently, she efficiently coordinated an entrepreneurship skills program at Pan African University, University of Ibadan, during the period 2020–2021. Professor Adepoju's expertise spans various areas, including re-skilling in the Fourth Industrial Revolution, Construction 4.0, Industry 4.0, human capacity development, human capital in renewable energy, assessing skill gaps to meet the demands of Construction 4.0, evaluating innovative skills, and management of Water, Energy and Food nexus.

Professor Nnamdi Ikechi Nwulu is a Professor of Sustainable Cyber-Physical Engineering Systems and Director of the Centre for Cyber Physical Food, Energy & Water Systems at the University of Johannesburg. Additionally, he is the Chairholder of the South Africa/Switzerland Bilateral Chair in Blockchain Technology. His research endeavors concentrate on the practical applications of digital technologies, mathematical optimization techniques, and machine learning algorithms within food, energy, and water systems.

Professor Nwulu's scholarly contributions encompass eight published books and over 70 indexed journal publications, earning him the prestigious 2021 NRF Research Excellence Award for Early Career/Emerging Researchers in the Engineering/Technology Category. He was also recognized as a finalist for the 2020/2021 TW Kambule-NSTF-South32 Award in the Emerging Researcher category and was honored with the 2020 UJ Vice Chancellor's Most Promising Young Researcher award. His memberships include the South African Young Academy of Science and various other professional associations. Furthermore, Professor Nwulu is the editor-in-chief of the *Journal of Digital Food Energy and Water Systems* and is Associate Editor for the *IET Renewable Power Generation* and *African Journal of Science, Technology, Innovation & Development.*

Dr. Love Opeyemi David is an interdisciplinary social scientist, proficient in project management, resource management, management science, digital transformation, and innovation management. He bagged first class honors in his bachelor of technology and earned a master of technology with distinction in project management technology from the Federal University of Technology, Akure, Nigeria. His academic journey culminated in a Ph.D. degree from the University of Johannesburg, South Africa.

Dr. David has showcased his expertise by publishing scholarly articles in Scopus-indexed journals. His diverse research interests encompass project management, innovation management, management science, Water-Energy-Food Nexus project delivery, circular economy, sustainable development/sustainability, and cutting-edge technologies of the Fourth Industrial Revolution (including blockchain technology, Internet of Things, and artificial Intelligence). He is currently the managing editor of the *Journal of Digital Food, Energy and Water Systems*.

Part I

Introduction and Background

1 Introduction

1.1 BACKGROUND TO THE STUDY

The breadth and depth of the Fourth Industrial Revolution have shown that all living and non-living things will be affected by the technologies of the revolution. The Fourth Industrial Revolution, also known as 4IR, brings a new dimension, opportunities, and predictable risks with high uncertainty of impact. Moreover, the extent and magnitude of its influence across the positive and negative dimensions are yet unknown, as there are endless combinations and tests of technologies of the Fourth Industrial Revolution used for different purposes. This level of uncertainty has caused academic fear and industrial tensions. Nevertheless, the Fourth Industrial Revolution and its technologies are an efficiency booster, a catalyst, a medium of technological discovery, and a ray of hope for an innovative breakthrough that will positively change the usual way of doing things in every aspect of life. Thus, the 4IR alters the pattern of getting things done and shifts the social, technological, and research equilibrium, necessitating a period of adjustment and the consequence of adjustment. Hence, there is a need for a framework to offer technological insights.

The Fourth Industrial Revolution combines economies of scale and economies of scope in generating a fusion of a new array of technologies, which, to a great extent, digitalized production process, whereby the process is monitored by the Internet of Things (IoT), cloud computing, and computers (Dombrowski and Wagner 2014). Postelnicu and Calea (2019) opined that the era of the Fourth Industrial Revolution is an era of digitalization, whereby digital technologies are applied to various fields of manufacturing, medicine, spatial missions, and aviation, among others, without compromising quality, which will lead to production cost minimization and higher economic reform. In describing how revolutionary, strategic, and dramatic the Fourth Industrial Revolution is, Schwab (2016) stated that the world is currently facing a shift across all industries, evident in new business models, changes in systems, and the reshaping of production mechanisms, consumption patterns, and transportation systems, whereby these changes are historical in terms of speed, size, and scope. The author further opined that the Fourth Industrial Revolution is a revolution for three reasons. The first reason is the velocity. Schwab (2016) opined that, unlike previous industrial revolutions, the Fourth Industrial Revolution is operating at an exponential rate rather than a linear pace because new technology is becoming more capable technology. The second reason is the breadth and depth of the revolution, which is not only changing the 'what' or 'how' of doing things but changing 'who' we are and redirecting our lifestyle and pattern. This has also led to a paradigm shift in business, individual lives, and the economy. The third reason is the impact of 4IR on systems, whereby an entire system is transformed across industries, companies, societies, and countries. Moreover, Deloitte Insights (2020) and David et al. (2022) posited

DOI: 10.1201/9781003522102-2

that the 4IR, also known as Industry 4.0, is the marriage between physical assets and advanced digital technologies such as artificial intelligence, drones, Internet of Things (IoT), robots, cloud computing, 3D printing, autonomous vehicle, and nanotechnology among others, which analyze, communicate, and act upon information that will enable consumer organizations and society to be more responsive and flexible to make more intelligent and data-driven decisions. The Fourth Industrial Revolution fuses the biological, physical, and digital world, combining new and emerging technologies and bringing humanity to globalization, whereby risks and opportunities are shaped (Mitrović 2020; Dogo et al. 2019).

Moreover, from the perspective of health, Castro e melo and Araujo (2020) opined that the 4IR is changing the dimension and system of the health sector in terms of the mechanism of understanding the health sector, transformation in the system, method, practice of diagnosis and administration of treatment, management style of the health system, and alteration of the relationship between patients and health professionals. The United Nations Industrial Development (2019) opined that the Fourth Industrial Revolution heralded a period where innovation is fast-paced, combining complex technology, making innovation more multidisciplinary, unplanned, collaborative, disruptive, and unpredictable. To justify that there is no universal definition of the 4IR, Soh and Connolly (2021) gave a unified description of the 4IR. The author opined that from the ontological point of view, the 4IR entails the interpretation of digital ecosystems with human lives and the environment. This also includes the broadened application of digitalization in economic activity, social dimensions, and political configuration. They argued that human bodies are becoming data, and there are no differences between the virtual world and reality due to digitalization. The second perspective of the 4IR description revealed that virtual and real-world interpenetrations will alter current and existing business models, whereby the supply chain is digitalized, trackable in real time, and automated, which will allow a reduction of waste, creation of new markets, and the tendency to lower prices of goods and services. The Department of Higher Education and Training (2020) opined that 4IR is changing in an exponential means the way humans create, distribute, and exchange value, which is resulting in a systemic change across different aspects of human life and sector with cross-cutting implications in social cycles, political dynamics, cultural mechanism, and economic pragmatism. The department further opined that the elements of the 4IR are divided into three along with their attributes, whereby this book divided it into four along with necessary modifications, as shown in Figure 1.1.

The Fourth Industrial Revolution has made remarkable impacts in several fields. Malomane et al. (2022) opined that through the application of the 4IR technologies in the construction industry, incidents of site injuries, and accidents could be curtailed, whereby construction sites are monitored and controlled, and the workforce can be trained virtually on safety and handling of equipment, ensuring accident prevention and the technologies can aid in better reporting, enhanced visibility, better communication, and improved workflows. Moreover, in the review of the 4IR, Deloitte (2020) stated that through the technologies of 4IR, there

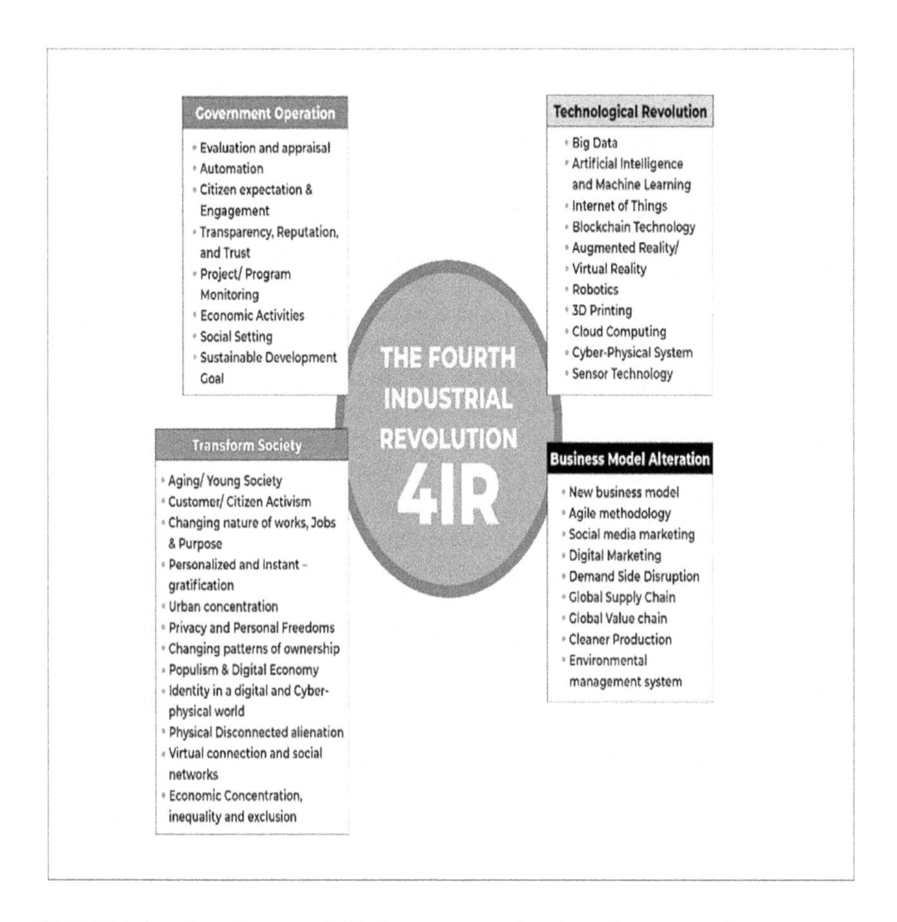

FIGURE 1.1 Classifications of 4IR (Department of Higher Education and Training 2020, p. 22)

would be disruptions in different organizations, especially in business strategies; the report noted that many organizations are utilizing big data analytics, artificial intelligence (AI), Internet of Things (IoT), and cloud computing to generate data for improved decisions-making, which will drive up revenue. UNIDO (2019) categorized the benefits of applying 4IR into three dimensions of sustainability. The economic benefits include enhanced control over the production process, lower transaction costs leading to higher revenue, competitive edge as a result of improved productivity, customer feedback mechanism leading to improved product, enhanced product quality, efficient management of global value chain processes in real-time across a great distance, a shift from mass production to mass customization, creation of intelligent production systems, conversion of customer behavior in creating sellable products, digital blend of manufacturing and service activities in boosting economic growth, and enhancement of predictive maintenance; 4IR acts as a catalyst to the United Nations Sustainable Development

Goals (SDG), to ensure leapfrogging in the advancement of some sectors of the economy and the safety and security of transaction data. Moreover, environmental benefits include the reduction of pollutants and emissions of greenhouse gases, providing greater resource efficiency, enabling accessibility and stability of natural resources like water and electricity; prevention, reduction, and elimination of waste; optimization of circular economy business model; capability to plan, analyze, predict, and control the performance of products; energy efficiency and energy savings; the capability of mitigating climate change; monitoring and tracking of environmental systems; and accelerating global climate agreements. Furthermore, the social benefits include enhancement of innovation and creativity, improvement in health, human cognition and physical capabilities, better working climate and workers' productivity, creation of opportunities for vulnerable and disadvantaged population groups, and creation of an integration path for SMEs to participate in global markets and ensure the transfer of knowledge.

Schwab (2015) also opined that the 4IR has the potential to improve quality of life, raise global income level, increase the ability of consumer access to the digital world, ensure a drop in the cost of communication and transportation, and enable the effectiveness of global supply chain and logistics. Goncharov (2020) analyzed the impact of the Fourth Industrial Revolution. Goncharov included automation of business, service, and production systems, quick penetration into the global market resulting from digitalization in research & development, sales, distribution and marketing, changing demands of goods and services, increased collaborative innovation, change in customer expectation and buying patterns, flexibility and sustainability of production assets, provisions of digital platforms in interaction and engagement between government and citizens, and restructuring of human's nature of privacy, consumer habits, accessibility, and changing of the dynamics to work and leisure. Gravier (2019) stated that the Fourth Industrial Revolution occurred in three stages and spheres, leading to several opportunities and risks. The first sphere is the digital-physical sphere, where real-world data, inputs, and innovations enter a digital sphere. The second stage is the digital sphere, where data and information in the digital sphere are processed into information, raising issues of security and integrity. The third sphere is the human sphere, which deals with the technical and digital skills needed in the integration process.

The Fourth Industrial Revolution is a digital enabler: redirecting, repurposing, and reconfiguring different sectors of the economy. For instance, it is changing strategic management decisions by influencing the decision-making process of senior executives and chief executive officers, who, before, had been subjected to long hours of brainstorming. It is also altering the administrative, business, and policy approach of tactical management, whereby middle-level managers rely on digital technologies occasioned by the Artificial Intelligence technology of 4IR in dissecting strategic goals and objectives. Also, the operation level of management utilizes the 4IR technologies in carrying out various tasks to meet organization outputs. In essence, it alters the organization's culture, vision, mission, objectives, and competitive strategy in building strategic agility.

Moreover, the educational system is not exempt from the Midas touch of the Fourth Industrial Revolution. It has changed the mode of learning, the dynamics of teaching, the medium of accessing knowledge, and the period of disintegrating knowledge. Moreso, the 4IR in the educational system, is changing the entire curriculum as a new way of doing things evolves, making old methods obsolete.

Furthermore, the Fourth Industrial Revolution is altering existing belief systems, whereby things, actions, and activities believed to be impracticable or impossible are becoming possible. Thus, changing societal norms and religious practices and altering value systems. The 4IR is also altering lifestyles, magnifying the imaginative capability of humans, boosting creative seeds, and indirectly stimulating the subconsciousness of human beings in their desires, wants, and needs. It is also making rural-urban migration quick but less interesting as, through the technologies, rural areas can access what urban dwellers enjoy. The 4IR is also altering the reproductive systems of humans, plants, and animals by determining gender, shape, nature, and metabolism, all of which impact the product of reproduction.

However, the Fourth Industrial Revolution has raised issues of the riskiness of the various technologies, which has raised palpable fears leading to reluctant adoption of the technologies, low exploration of the technologies, and avoidance of the technologies. The Technology Acceptance Model of Fred Davis (1986) as collaborated in David et al. (2022) has shown that the acceptability of a new array of technologies is not automatic but based on individual subjectivity, perceived usefulness, perceived ease of use, relevancy of job, quality of output, and practicability of results, which will influence the intention of an individual or an organization in using such technology. However, the depth and breadth of the perceived ease of use always bring about the nature of the risk involved. This is also in conformity with the Innovation Diffusion theory of Rogers (2003), which explains that the rate of spread and acceptance of innovation, relative advantage, compatibility, complexity, trialability, and observability all determine the adoption of the innovation, whereby adopters are segregated into various period of adoption. This smirks of the fear of risks. Therefore, based on the Technology Acceptance Model (TAM) and Innovation Diffusion Theory (IDT), identifying and mitigating risks for ease of adoption and effectiveness of spread for the Fourth Industrial Revolution necessitates this book.

Risk itself is unavoidable in the era of the 4IR. Still, managing risks is a preventive and mitigating strategy in optimizing the benefits of the 4IR technologies and avoiding a closed loop of bad and good, which may affect not just the outcome of 4IR but the entire human race. However, risks mean different things to different individuals and organizations. Risk is relative, as what is risky to party A may not be risky to party B. This also goes for individual attitudes toward risks. Outreville's (1998) research gave different definitions and descriptions of risks. They include the following: "it is defined as uncertainty to loss"; "it is the uncertainty of financial loss"; "it is defined as the perils to which an individual is objectively exposed at any time"; "it is defined as the variation in the outcomes that could occur over a specified period"; "it is defined as the lack of predictability of outcome"; and "it is defined as a condition in which there is a possibility of an adverse deviation from the desired outcome." However, from the perspective of this book about the

Fourth Industrial Revolution, the risk is an outcome that may adversely affect an intended or unintended process, product, result, or outcome during the utilization of one or more technologies of the Fourth Industrial Revolution.

Several types of research have identified risks and potential risks in utilizing the technologies of the Fourth Industrial Revolution. The study by Adepoju et al. (2022) averred that the technologies of the Fourth Industrial Revolution would lead to the automation of skills, leading to a high unemployment rate, with various hazardous effects on society. The author also opined that through the advert of 4IR, crime would be nearly impossible to control due to the sophistication of 4IR technologies, which will be rampant among drug cartels, non-state actors, fraudsters, and terrorists. The author opined that through nanotechnology, mass production of weapons which are not easily detectable could lead to the mass destruction of lives and properties. Castro et al. (2020) opined that the technologies of 4IR can lead to a reduction in human relations, which has profound health implications. Also, Postelnicu and Calea (2019), while acknowledging the creation of new jobs due to the technologies of 4IR, averred that job loss is inevitable, as the World Economic Forum (2016) stated that due to the technologies of 4IR, labor disruption would lead to about 5.1 million job losses, while many professions will become redundant. Schwab (2016) also shared this perspective, stating that the Fourth Industrial Revolution will create fewer jobs, unlike the previous revolution. Piggin (2016) opined that the 4IR technologies constitute a significant risk to information technology (IT) systems as there are invasions and intrusions into control systems, thus subjecting the IT systems to manipulation, corruption of data, and data insecurity.

This research shows a consensus that the 4IR technologies exhibit risks to people's socioeconomic dimensions. The study of Mitrović (2020), Castro E melo and Arauyo (2020), Postelnicu and Calea (2019), World Economic Forum (2016), Schwab (2016), Piggin (2016), and Zervoudi (2020) all showed the general risks of the Fourth Industrial Revolution without an in-depth analysis of the risks to each of the technologies of the Fourth Industrial Revolution. Also, there is a lack of research on the risks attached to the 4IR technologies as a compendium of knowledge or research publication. Therefore, this book bridges this intellectual and technological lacuna by highlighting and explaining the various risks that can occur from the technologies of the Fourth Industrial Revolution. Furthermore, there is a shortage of risk management approaches, methodologies, and frameworks for the Fourth Industrial Revolution technologies. This book answers this intellectual curiosity on managing the risks of each of the significant technologies of the Fourth Industrial Revolution.

1.2 AIM AND OBJECTIVES

This book aims to assess and provide a risk management framework for the technologies of the Fourth Industrial Revolution. The objective of this book includes an assessment of risk in the Fourth Industrial Revolution, the impact of Fourth Industrial Revolution Risks on the people, industries, and economy, current risk management techniques, an examination of the various risks associated with the Fourth Industrial Revolution technologies and risk management skills for each of the technologies such as artificial intelligence, Internet of Things (IoT), blockchain technology, cloud computing, robotics, and augmented reality.

1.3 ORGANIZATION OF THE BOOK

The book is divided into three parts. Part I contains three chapters, of which Chapter 1 is the introduction to the book, constituting the background to the study, aim and objectives, organization of the book, and contribution and values of the book. Chapter 2 introduces the Fourth Industrial Revolution along with the various 4IR technologies, an overview of risks in the Fourth Industrial Revolution, categories of risks, and the impact of risks on the Fourth Industrial Revolution. Chapter 3 contains the risk identification process, risk response planning and techniques, quantitative and qualitative risk analysis, and risk monitoring and controlling. Part II includes six chapters, each addressing technologies of the 4IR: artificial intelligence, the Internet of Things, blockchain technology, cloud computing, robotics, and augmented reality. Each of the six chapters contains the overview and components of the technology, risks associated with the technology, risk management technique, and risk management skills for the technology. Part III has only one chapter, which is the risk management framework for the Fourth Industrial Revolution, constituting the risks of the 4IR, legal, political, and socio-environmental risks of 4IR, risk mitigation measures, and the 4IR risk management.

1.4 CONTRIBUTION AND VALUE

The book contributes to the pool of resources on 4IR, with a distinctive contribution to an area not yet covered, which is a compendium of 4IR risks and how to manage them. The book will be an intellectual legacy on risks and risk management of the Fourth Industrial Revolution and the individual technologies of the Fourth Industrial Revolution.

Moreover, it provides valuable insights into risks for all professions, as the technologies of 4IR cover all professions of human endeavor. Its value is a technological guide for IT personnel and management staff during the production process to watch out for what needs to be avoided and what needs planning.

1.5 SUMMARY

The Fourth Industrial Revolution technologies are technologies reconfiguring the world and redirecting methodologies and production mechanisms. Hence, risks are bound to happen in the process of utilization. Therefore, this chapter provides an overview of the expectations of the book's content, which is the first book on risk management framework of 4IR technologies. It assesses the background of the book, the challenges that stimulate the solution provided by this book, the gap covered by this book, and the contribution of the book.

REFERENCES

Adepoju, O., Aigbavboa, C., Nwulu, N., and Olaiya, M. (2022). *Reskilling Human Resources for Construction 4.0. Implications for Industry, Academia, and Government.* Springer. https://doi.org/10.1007/978-3-030-85973-2.

Castro, E.M., António, J., De Melo, E.G., and Faria Araújo, N.M. (2020). *Impact of the Fourth Industrial Revolution on the Health Sector: A Qualitative Study.* The Korean Society of Medical Informatics. https://doi.org/10.4258/hir.2020.26.4.328.

David, L.O., Nwulu, N.I., Aigbavboa, C.O., and Adepoju, O.O. (2022). Integrating Fourth Industrial Revolution (4IR) Technologies into the Water, Energy & Food Nexus for Sustainable Security: A Bibliometric Analysis. *Journal of Cleaner Production* 363. https://doi.org/10.1016/j.jclepro.2022.132522.

Deloitte Insights. (2020). *The Fourth Industrial Revolution at the Intersection of Readiness and Responsibility.* London, United Kingdom: Deloitte Insights.

Department of Higher Education and Training. (2020). *Report of the Ministerial Task Team (MTT) on 4IR for the Post-School Education and Training System.* Pretoria: Department of Higher Education and Training.

Dogo, E.M., Salami, A.F., Nwulu, N.I., and Aigbavboa, C.O. (2019). Blockchain and Internet of Things-Based Technologies for Intelligent Water Management System. In: Al-Turjman, F. (ed.) *Artificial Intelligence in IoT: Transactions on Computational Science and Computational Intelligence.* Cham: Springer. https://doi.org/10.1007/978-3-030-04110-6_7.

Dombrowski, U., and Wagner, T. (2014). Mental Strain as Field of Action in the 4th Industrial Revolution. Variety Management in Manufacturing. *Proceedings of the 47th CIRP Conference on Manufacturing Systems.* Procedia CIRP 17 2014, pp. 100–105. https://doi.org/10.1016/j.procir.2014.01.077.

Goncharov, V.V. (2020). The Fourth Industrial Revolution: Challenges, Risks and Opportunities. *Eruditio e-Journal of the World Academy of Art & Science* 2(6), 95–106. https://eruditio.worldacademy.org/files/vol2issue6/reprints/The-Fourth-Industrial-Revolution-VGoncharov-Eruditio-V2-I6-Reprint.pdf

Gravier, M. (2019). *Risks and Opportunities in the Fourth Industrial Revolution. Supply Chain Management Review.* Peerless Media LLC. https://www.scmr.com/article/risks_and_opportunities_in_the_fourth_industrial_revolution/management.

Malomane, R., Musonda, I., and Okoro, C. S. (2022). *The Opportunities and Challenges Associated with the Implementation of Fourth Industrial Revolution Technologies to Manage Health and Safety.* MDPI AG. https://doi.org/10.3390/ijerph19020846.

Mitrović, L.M. (2020). *Challenges, Risks and Threats to Human Security in the 4th Industrial Revolution.* Centre for Evaluation in Education and Science (CEON/CEES). https://doi.org/10.5937/nabepo25-26316.

Outreville, J.F. (1998). The Meaning of Risk. In: *Theory and Practice of Insurance* (pp. 1–12). Boston, MA: Springer US.

Piggin, R. (2016). Risk in the Fourth Industrial Revolution. *ITNow* 58(3), 34–35.

Postelnicu, C., and Calea, S. (2019). The Fourth Industrial Revolution, Global Risks, Local Challenges for Employment. *Montenegrin Journal of Economics* 2, 195–256. https://doi.org/10.14254/1800-5845/2019.15-2.15.

Rogers, M.E. (2003). *Diffusion of Innovations.* 5th Edition, New York: Free Pass.

Schwab, K. (2015). The Fourth Industrial Revolution. What It Means and How to Respond. *SNAPSHOT.* https://cdn.lgseta.co.za/resources/research_and_reports/4IR%20Resources/The%204IR_What%20it%20Means%20and%20How%20to%20Respond_Klaus%20Schwab_2015.pdf.

Schwab, K. (2016). *The Fourth Industrial Revolution.* Cologny: World Economic Forum.

Soh, C., and Connolly, D. (2021). New Frontiers of Profit and Risk: The Fourth Industrial Revolution's Impact on Business and Human Rights. *New Political Economy* 26(1), 168–185.

United Nations Industrial Development Organization, UNIDO. (2019). *Bracing for the New Industrial Revolution Elements of a Strategic Response Discussion Paper*. Vienna.

Zervoudi, E.K. (2020). *Fourth Industrial Revolution: Opportunities, Challenges, and Proposed Policies*. IntechOpen. https://doi.org/10.5772/intechopen.90412.

2 The Fourth Industrial Revolution

2.1 INTRODUCTION TO THE FOURTH INDUSTRIAL REVOLUTION

Technological revolutions have always been built on preceding revolutions; however, there are four major industrial revolutions, of which the world is currently experiencing the Fourth Industrial Revolution. According to Xu et al. (2018), Klaus Schwab, the founder and executive chairman of the World Economic Forum, coined the term "Fourth Industrial Revolution." Like the preceding three revolutions, the fourth industrial revolution entails advancing science and technology, changing production patterns, machine utilization, and the technological revolution of industrial activities. The Natural Academy of Science and Engineering (2013), Liao et al. (2018), and David et al. (2022) explained the four industrial revolutions. The first industrial revolution started in the 18th century, when water and steam-powered production mechanization led to production efficiency having eight times better productivity. During the 1st Industrial Revolution, the steam engine enabled the transition from the traditional farming system under a feudal society to new production methods, where coal was the main energy and steam-powered trains were the means of transportation. The second revolution was characterized by the invention of the internal combustion engine between the end of the 19th century and early 20th century, leading to electrically powered mass production of goods with optimized speeds and minimized costs. The third revolution began in the mid-20th century, popularizing electronics and information technology to automate production using computers and programmable controls. The Fourth Industrial Revolution is the current revolution permeating all human endeavors. It is characterized by the technologies of information and communication concepts, permitting operations between the cyber and physical systems. According to Prisecaru (2016), cited by Xu et al. (2018), Table 2.1 summarizes the four industrial revolutions.

According to Prisecaru (2016), many high-profile business leaders, technology enthusiasts, and world leaders have described the Fourth Industrial Revolution differently. Schwab, the World Economic Forum (WEF) chairman, opined that the Fourth Industrial Revolution "is characterized by a much more ubiquitous and mobile Internet, by smaller and more powerful sensors that become cheaper, and by artificial intelligence, and machine learning, and one may see its evolution in a world, in which virtual and physical human systems are intertwined in services, manufacturing and other human endeavors." Prisecaru (2016) further posited that the Fourth Industrial Revolution would enable industrial development, sustainable solutions, reduction of industrial waste, and redesigning of production and

DOI: 10.1201/9781003522102-3

TABLE 2.1
Summary of the four industrial revolutions

Revolution	Period	Transition Period	Energy Resource	Main Technological Achievement	Main Industries	Transport Means
1st	1760–1900	1860–1900	Coal	Steam engine	Textile, steel	Train
2nd	1900–1960	1940–1960	Oil, electricity	Internal combustion engine	Metallurgy, auto, machine building.	Train, car
3rd	1960–2000	1960–2000	Nuclear energy, natural gas	Computers, robots	Auto, chemistry	Cars, planes
4th	2000 till date	2000–2010	Green energies	Internet of Things, 3D printer, genetic engineering.	High technology industries	Electric cars, ultra-fast trains

Source: Prisecaru (2016)

consumption system, accelerate the cycle of innovation, introduce news requirement for the educational system, enable a broad disruption of the labor market, cause a change in the fiscal policy, and help in reaching global goals.

Stancioiu (2017) averred that the Fourth Industrial Revolution entails technological revolutions in various fields, including information and communication technology applications for digitalizing and integrating information systems in design and manufacturing and new software for simulation modeling, digital manufacturing, and virtualization. Other applications include developing cyber-physical systems for monitoring and controlling physical processes, simplifying manufacturing processes using 3D printers and additive manufacturing, and new forms of human-machine interactions. The author noted that the benefits of the Fourth Industrial Revolution encompass efficiency and accuracy in time, cost, the flexibility of systems, and the integration of various processes in making informed decisions. According to Dechprom and Jermsittiparsert (2018), the Fourth Industrial Revolution has shifted countries' economies of scale from the supply-side economy to the demand-side economies of scale, thereby leveraging technological innovations to reduce marginal costs on the supply side to zero.

Schwab (2016) stated that the Fourth Industrial Revolution amalgamated the physical and virtual manufacturing systems to customize products and new production models. The author further opined that the Fourth Industrial Revolution entails smart machines and systems, technological breakthroughs in gene sequencing and nanotechnology, and the interactions between physical, digital, and biological spaces. The author sees the Fourth Industrial Revolution as an optimizer of the 2nd and 3rd revolutions. About 1.3 billion people still lack access to electricity, and about four million people do not have Internet access. Schwab (2016) stated that the Fourth Industrial Revolution (4IR) had the following impacts: a surge in high productivity and economic growth, positively affecting aging for people to live long, the possibility of technological unemployment, aid labor substitution, up-skilling and reskilling of skills, reshoring of global manufacturing systems and processes, alteration like work, assets productivity, a shift in customer expectations, and new forms of collaborative innovation.

The Fourth Industrial Revolution encompasses a cyber-physical system, which entails the integration of sustainability, production, and customer satisfaction, forming the basis of intelligent processes and network systems (Bloem et al. 2014; Dogo et al. 2019). It is the technological integration of cyber-physical systems into logistics, manufacturing, and the Internet of Things in industrial processes (National Academy of Science and Engineering 2013). Ruzavorsky et al. (2020) opined that the Fourth Industrial Revolution, also known as Industry 4.0, is the digitalization of industrial production, leading to a fully interconnected, intelligent, and digitalized organization of activities and production factories. It is revolutionizing production from dominant machine manufacturing to digital manufacturing (Oztemel and Gursev 2018). Lu (2017) avowed an integrated, adapted, and interoperable manufacturing process in which big data, algorithms, and emerging technologies are included in transforming production patterns from mass production to mass customization. Poljak (2018) noted that the Fourth Industrial Revolution integrates Internet control systems with existing normalities to allow a seamless connection between people and machines anywhere, anytime, and with anything, irrespective of the complexity. Industry 4.0, according to Lee et al. (2018), is affecting all types of industries in the primary sector (industries involved in the extraction and production of raw materials), secondary sector (industries involving the conversion of raw materials to finished products), and tertiary sector (the service sector of the economy).

The Fourth Industrial Revolution is now necessary for the rapid development of many endeavors, as it ensures the efficiency of activities, digitalization of processes, and data accuracy, resulting in longevity. It has permeated all aspects of human endeavors, bringing together the nexus of different digital, biological, and physical spaces. The Fourth Industrial Revolution is the progress of preceding revolution outcomes, as it provides a more strategic way of actualizing the outcomes. It is an evolving technological revolution, as current research has shown that it is an explosion of technological innovations, giving speed to actualizing humanity's mental capability of ideas and thoughts. The Fourth Industrial Revolution is beyond the digitalization of activities or the production of intelligent products. It is a shift in human existence, as these technological outcomes shape humanity's purpose and fulfill biblical assertions of replenishing the earth. The Fourth Industrial Revolution is the technological exhibition and the exploration of the wonders of the human mind, which can be catastrophic or an agent of positivity, highlighting the risk aspect of Industry 4.0. The Fourth Industrial Revolution is an intelligent stimulator and digital integrator of the wonders of the universe and wonders of the human mind, leading to the proliferation of different innovations and creativities. The Fourth Industrial Revolution is like a universal code and technological captor, shaping the foods we eat, the type of place we live in, our mode of transportation, the nature of our health system, how we earn and spend money, our communication system, altering business modus operandi, regulating human laws, altering our justice system, shaping our educational system and mode of learning and more importantly, making all our activities transparent.

Some technologies that make up the Fourth Industrial Revolution are different in applications and sometimes share the same area of applications. Ruzarovsky et al. (2020) and Adepoju et al. (2022) opined that the Fourth Industrial Revolution is premised on nine main technologies, which are system integration, additive manufacturing, autonomous robots, 3D printing, augmented reality, cloud computing, big data, cyber security, cyber security, Internet of Things, and simulation. Koh et al. (2019) stated that the technologies are grouped into two: five main technologies and five emerging technologies. The main technologies are the Internet of Things (IoT), cloud computing, big data analytics, 3D printing/additive manufacturing, and robotic systems. Digital twin technology, blockchain, machine learning, artificial intelligence, and 5G technologies are emerging technologies. According to Liao et al. (2018), while researching public policies on the Fourth Industrial Revolution, there are 10 prioritized technologies in public policies. The technologies are advanced robotics, big data analytics, cloud computing, cognitive computing, cyber security, the Internet of Things, machine-to-machine, mobile technologies, 3D printing, and RFID technologies.

Major technologies of the Fourth Industrial Revolution have permeated all spheres of human endeavor and directly affected humans. They include the Internet of Things (IoT), artificial intelligence, blockchain technologies, cloud computing, robotic technologies, and augmented/virtual realities. These technologies cut across the leading and emerging technologies used in public and private spaces. This chapter briefly introduces these technologies, while detailed analyses and explanations are given in each succeeding chapter.

2.1.1 Internet of Things

The Internet of Things (IoT), from its name, connotes interaction, interoperability, and communications among all kinds of things. Stancioiu (2017) refers to the Internet of Things as the infrastructure of the information society, which entails the interconnection of intelligent components through a network for collecting and exchanging information. The IoT is the network of physical objects with embedded technology for interaction and communication with both internal and external environments (Karabegovic and Husak 2018). Patel and Patel (2016) also opined that the Internet of Things is an ecosystem of the Internet, through which there is a connection between people and things, things and things, and people and people. The author further posited that IoT is the foundational technology of smart energy, living, industry, smart cities, smart health, smart buildings, smart transport, and smart homes. According to Mohammed and Ahmed (2017), objects and things recognize themselves through IoT and get intelligent information, behavior, or patterns in making decisions. The author summarized the Internet of Things (see Figure 2.1).

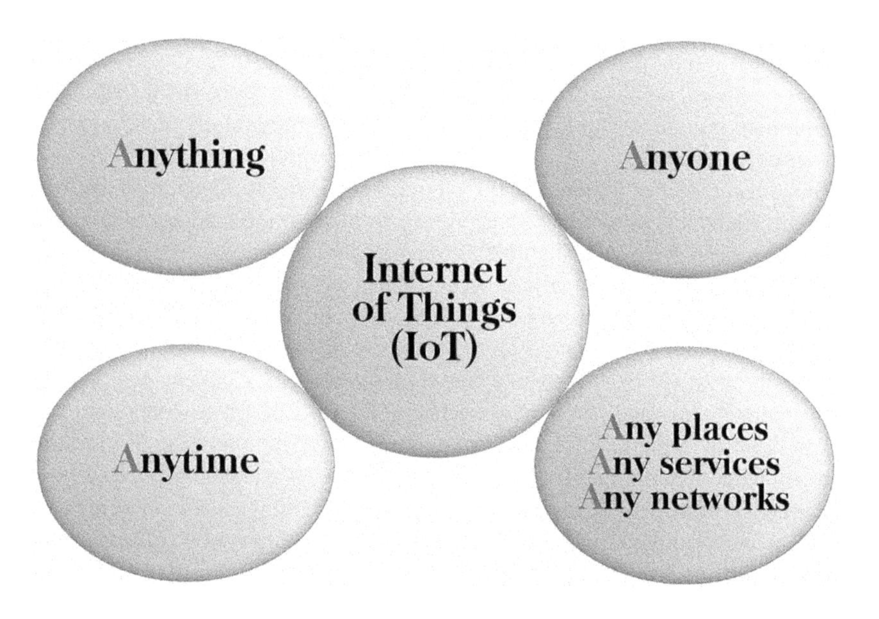

FIGURE 2.1 Internet of Things (Mohammed and Ahmed 2017, p. 127)

2.1.2 Artificial Intelligence

According to John McCarthy, often regarded as the father of artificial intelligence, artificial intelligence "is the science and engineering of making intelligent machines, especially intelligent computer programs" (McCarthy 2007). According to the author, AI entails the creation of machines that will solve problems by observing and perfecting the mechanism of human intelligence. It is developing and creating computer systems, either hardware or software, to perform tasks requiring human intelligence. It is simply the computerization of human intelligence in solving problems or performing activities in any field of human endeavors. According to TutorialPoints (2015), artificial intelligence is the mechanism of making computer system, or applications think intelligently like a human being, intending to create an expert system (i.e., a system that exhibits intelligent behavior, demonstrates, learns, explains and advises its users/operators) and implementation of human intelligence in machines (a system that can think, understand, behave, and learn like humans). According to the organization, artificial intelligence has been applied in games, natural language processing, experts' system, vision/virtual systems, speech recognition, handwriting recognition, and intelligent robots. AI systems comprise different forms of intelligence, including linguistic intelligence, logical-mathematical intelligence, spatial intelligence, kinesthetic intelligence, interpersonal intelligence, intrapersonal intelligence, learning intelligence, physiology intelligence, and philosophical intelligence.

2.1.3 Blockchain Technology

The US National Institute of Standards and Technology opined that blockchain technology is a modern cryptocurrency product, which is a tamper-evident and resistant digital ledger implemented in a distributed mechanism and without a central authority (Yaga et al. 2018). It is a technology for transactions without regulations or centralization but standardization. Zheng et al. (2017) stated that blockchain technology (one of the technologies of the Fourth Industrial Revolution) is a public ledger where all committed transactions are stored in a list of blocks where chains of blocks grow as new blocks and are appended continuously. According to the authors, the characteristics of the blockchain technology are as follows: decentralization (this means that there is no centralized mode of operation, as there is a consensus algorithm used in maintaining data consistency in a distributed network) and persistency (i.e., quick validation of transactions, whereby invalid transactions are discovered immediately). Other characteristics are anonymity (the real identity of users can't be revealed) and auditability (transactions are stored in an unspent transaction output, UTX—O model, where unspent transactions are regularized). Ahram et al. (2017) opined that according to Wild et al. (2015), blockchain technology is defined "as a network of computers, all of which must approve a transaction that has taken place before it is recorded, in a chain of computer code. The transfer details are recorded on a public ledger that anyone on the network can see." The uniqueness of blockchain technology lies in its modus operandi, which is transparent and self-accountable to the user without any system regulating it. Hence, Holotescu et al. (2018) stated that blockchain technology establishes a decentralized technological situation where the cryptographically authenticated transactions and data are not within the control of any third party. Also, finalized transactions are recorded in an unassailable ledger, which is certifiable, transparent, sheltered, and permanent with a timestamp and other details.

2.1.4 Cloud Computing

The US National Institute of Standards & Technology (NIST) defines cloud computing "as a model for enabling convenient, ubiquitous, on-demand network access to a shared pool of configurable computing resources (such as networks, servers, services, storage, and applications), which can be released with minimal management effort or service provider interaction" (Srinivas et al. 2012; Rashid and Chaturvedi 2019). Rashid and Chaturvedi (2019) averred that cloud computing simply stores and accesses data, information, and programs over the Internet instead of physical computer storage devices. The authors posited that cloud computing is a system that contains infrastructure, software, and a platform, with the characteristics of cost-effectiveness, on-demand self-service, broad network access, resource pooling, rapid elasticity, measured services, multitenancy, scalability, reliability, efficient resource utilization, and virtualization. Typically, cloud computing is a technology that allows access to cloud-stored information and data anywhere and at any time, irrespective of your geographical location. This cloud storage is enabled by Internet capability, sometimes called Internet storage. It is a technology that acts as information and is regulated by different parties/organizations/platforms based on acceptable standards and service differentiation.

2.1.5 Robotic Technology

Robotic technology entails manufacturing robots, which Williams (2021) defined as electromechanical devices with multiple degrees-of-freedom (dof), which are programmable to accomplish various tasks. Robotic technology is designing a specialized programmable computer system or devices with specific tasks, instructors, and activities. According to Umachandran (2020), robotic technology is an interdisciplinary field of computer science, information engineering, and electromechatronics to compute and control through sensory feedback and data in supporting activities and applications that substitute human actions. It replaces man in certain activities for efficiency, effectiveness, accuracy, and reprogrammable-oriented data storage. The Robot Institutes of America) defined a robot as a "reprogrammable multifunctional manipulator designed to move materials, tools, parts or specialized devices through variable programmed motions for the performance of varieties of tasks, which also acquire information from the environment and more intelligent in response" (Hegel et al. 2009). Robotic technology is a fusion of artificial intelligence applications, the architecture of the Internet of Things, and possession of the storage capacity of cloud computing for the performance of specific tasks. It could also be viewed as the end product of the Fourth Industrial Revolution.

2.1.6 Augmented Reality

The clarity of virtualization, the virtual presence of a person in a virtual world, the prescriptive capability of virtual representation, and the predictive abilities of virtual information make augmented reality a cutting-edge technology of the Fourth Industrial Revolution. Silva et al. (2003), Bottani and Vignali (2019), Abdulhameed et al. (2019), and Chen et al. (2019) opined that augmented reality is a technology augmenting virtual reality and telepresence, where users see the real world augmented with virtual objects. According to the authors, virtual reality comprises a computer-generated 3D environment, allowing a person to enter and interact with a synthetic environment in the artificial computer world. In contrast, telepresence entails extending an individual's sensory-motor facilities and problem-solving abilities in a remote environment. Augmented reality combines real and virtual worlds, real-time interaction, and 3D registration (Silva et al. 2003; Carmigniani and Furht 2011). Chen et al. (2019) opined that the technicalities of AR include 3D-modeling, multimedia, intelligent interaction, real-time tracking and registration, sensory, and more, where it simulates the real world with virtual computer information like music, video, 3D models, text, and images.

2.2 OVERVIEW OF RISKS IN THE FOURTH INDUSTRIAL REVOLUTION

The Fourth Industrial Revolution is the technological revolution permeating all aspects of human endeavors and cutting across disciplines. However, changing the dynamics of technological realities in the world has adverse consequences. The resistance to change, adaptation, and fusion of several changes resulting from the Fourth Industrial Revolution all come with risks. There is a need to critically review the socio-technological configuration and risks resulting from the Fourth Industrial Revolution.

According to the American Risk and Insurance Association, risk is "uncertainty as to the outcome of an event, when two or more possibilities exists" (Outreville 1998). According to the author, these risks could be static (risks that involve losses that would occur even when the economy does not change) or dynamic risks (involved in the possibility of a dynamic change). It could be either speculative risks (where there is a possibility of a loss or gain) or pure risks (the prospect of a loss); or particular risks (risks that arise out of individual action and are only felt by individual and not group); or fundamental risks (risks that affect a large group of people).

Fischhoff et al. (1984) opined that the risks associated with technology have led to both the development and the demise of some industries, empowering the powers of some agencies and overwhelming the capacity of some agencies, enhancing the growth of multidisciplinary fields and distorting the growth of other disciplines. The research of Renn and Benighaus (2013) defined technological risks as "the likelihood of physical, social and financial harm/detriment/loss as a consequence of a technology aggregated over its entire lifecycle." According to the author, technological risks are often used interchangeably with technological hazard, which is the potential threat or harm from technology or a product of technology to people, capital, nature, or human-made facilities or projects.

Risks arising from the technologies of the Fourth Industrial Revolution are unprecedented and unavoidable. These risks arising from the fusion of these technologies will lead to a tiny weapon of mass destruction, accidental deaths of careless or carefree users or operators, altering world power balances and putting humanity at its knees. The development and configuration of pathogens and viruses from the technologies associated with synthetic biology (designing and constructing biological equipment, tools and systems) make technological risks scary. The rate of deaths related to viruses and diseases, which can be engineered from synthetic biology, makes technological risks of the Fourth Industrial Revolution a global discourse. The technological capability to alter the world's climatic system or even a region is a technological hazard in waiting. Risks associated with the Internet of Things give global access through the Internet gate to anyone to control and is a technological risk that can alter the lifestyle, value systems, safety and security, and shelter of man, either positively or negatively, depending on the prevailing global interests.

The research of Karjalainen et al. (2019) on the Fourth Industrial Revolution risks along the circular economy and environment observed the following risks: bias in technology, invisibly and misjudging the potentials of emerging technologies, issues relating to privacy, mental health disruption due to the establishment of a new balance between society and self, weakening of the connections in nature, and cost of technological solutions. Onik et al. (2019) explored the risks associated with personal data privacy concerning the technologies of the Fourth Industrial Revolution. Artificial intelligence and robotics technologies exhibit the risks of data privacy issues, which are a lack of privacy standardization for AI technologies and inadequacies in monitoring AI decision-making regarding profiling to avoid data corruption. According to the authors, using cookies and/or beacons is a privacy risk issue associated with augmented reality (AR) and virtual reality (VR). Data leakage is a risk in most Fourth Industrial Revolution technologies due to encryption issues, raw data storage, and insufficient standardization organization.

The risks of the Fourth Industrial Revolution are risks that are embedded with the technologies. They are risks that manifest upon utilization, reconfiguration, adaptation, and integration with other technologies or projects involving the technologies. These uncertain possibilities, sometimes hazardous, are the focal points of this book, having a significant impact on humanity's sustainability drive, especially when the desire for technological breakthrough overrides the systematic process of detecting potential issues or related laws governing the usage process.

2.3 CATEGORIES OF RISKS IN THE FOURTH INDUSTRIAL REVOLUTION

Risk occurrences associated with any of the technologies of the Fourth Industrial Revolution are risks that have a devastating effect on the livelihood of human beings, the progress of humanity, the gains of globalization, the prospects of economic boom, the technological breakthrough of emerging technologies, and the fragility of peace among the countries of the world.

The risks of the Fourth Industrial Revolution are categorized into the following:

1. Economic risks
2. Human risks
3. Social risks
4. Environmental risks

2.3.1 Economic Risks

The Fourth Industrial Revolution is an economic booster, and stabilizer and offers optimal and efficient business solutions. According to a PWC survey in 2020, 63% of business leaders stated that the 4IR would provide protection against economic downturns through an increase in the value of their: products and services, the creation of new streams of revenue, expansion of the economic benefits of globalization, aid localization and reduction of labor costs. However, Soh and Connolly (2021) opined that the Fourth Industrial Revolution would have economic consequences. There is an economic opportunity cost of the benefits of the Fourth Industrial Revolution on the economy. This is because the promotion of technologies furthers an economic interest, which is more of a proliferation of socio-technical visions. Economic risks from the Fourth Industrial Revolution include job losses, income inequality, worker's exploitation, dangerous working environment due to the technologies of 4IR, new business models of prosumerism and gamification, technological abuse of human rights in business, and the manipulation of the emotions for profit-making (Soh and Connolly 2021). The research of Zervoudi (2021) stated that economic risk from the Fourth Industrial Revolution technologies is the risk of automation of jobs, where there is a prediction of job automation in 57% of jobs in OECD countries, 47% of jobs in the United States, and 77% jobs in China. Another economic risk is the widening income inequality gap, especially due to the high number of low-skilled and inexperienced workers and the exposure of automation to jobs.

2.3.2 Human Risks

Human beings are the users, operators, and custodians of the Fourth Industrial Revolution technologies and are very susceptible to the thorns and daggers of revolutionary technologies. According to Brochal et al. (2019), the risks affecting humans are usually linked to and occupation risks. Most risks that affect human beings are usually fundamental risks. The risks of human-machine interaction and human-robot interaction can lead to industrial accidents and cause severe health issues. Human exposure to technologies that need personal protective equipment (PPE) is susceptible to exposing their bodies to harm, especially in underdeveloped countries with low PPE usage. Mitrovic (2020) opined that the technologies of the Fourth Industrial Revolution would breach human security, which the United Nations Development Program (UNDP) report of (1994) defined as "freedom from fear" and freedom from want." According to the authors, the Fourth Industrial Revolution will fuel, navigate, and spread the existing security issues occasioned by globalization. These securities issues are from the seven dimensions of human security: economic security (unemployment, homelessness, poverty and inflation, among others), environmental security (natural disasters, ecosystem degradation, etc.), health security (parasitic disease, viruses, infectious diseases), personal security (workplace violence); community security (ethnic tension, violent conflicts, etc.), food security (economic access to healthy food), and political security (human rights violations). Other risks that affect humans, as elucidated by Mitrovic (2020), include the exportation of unsafe products, human being trafficking, weapons proliferation and arms trafficking, industrial espionage, white-collar crimes, increase in the scope and diversity of crimes.

There are human risks that are attached to individual technologies of the revolution. For instance, nanotechnology and 3D printing technologies aid the production of nano-weapons and create a means of avoiding weapon control. Also, the application of synthetic biology in the pharmaceutical industry enhances the mass production of terrorist drugs known as amphetamine and phenethylamine, which boosts suicide bomber's confidence. Moreover, technologies in biotechnology using genetic engineering create the possibility and rapid spread of new deadly viruses. The Internet of Things technologies and devices make a person vulnerable to attacks of any magnitude due to the possibility of hacking stored data about one's lifestyle and daily activities. The Fourth Industrial Revolution is emerging as an epicenter of dynamic change, altering the lifestyle of human beings, leading to mental redundancy and reduction of human creativity. This results from the coming impact of intelligent machines, occasioned by artificial intelligence technologies, requiring minimal human effort. This could also lead to the risk of a mental health crisis.

Exposing on the health risks for human beings occasioned by the Fourth Industrial Revolution, Min et al. (2019) opined that the Fourth Industrial Revolution is causing a change of dynamics in an organization in terms of work automation and shift of work. Risks identified in these changes include the circadian rhythm disturbance by change of melatom (difficulties in sleeping) resulting from shift work. Also, shift work, which 4IR may occasion, leads to breast cancer, colon cancer, and prostate cancer. Working in shifts also leads to the risk of cardiovascular and cerebrovascular diseases, coronary heart disease, ischemic stroke, diabetes, obesity, and the incidence of depression.

2.3.3 Social Risks

Social sustainability should be the most significant outcome of the Fourth Industrial Revolution, as it encompasses the nature and nurture of humanity. However, the Fourth Industrial Revolution poses some social risks to humans, given the dependency traits of the technologies of 4IR. Workers in a typical organization tend to lose their identity and self-esteem when intelligent machines replace their functionality or uniqueness in the form of robots. Also, migration is possible due to the activities emanating from 4IR. This may result in social risks of a language barrier, exploitation of racism, and difficulties in assessing basic social amenities. The creation of new jobs from the Fourth Industrial Revolution encompasses the social risks of no distinction between private life and work life, skill mismatch, and politicization of the workplace.

Moreover, research has shown a dichotomy regarding technological skills between men and women. However, the advent of the Fourth Industrial Revolution brings out the risks of a widening technological gap between men and women, aggravating gender inequality. Also, most jobs favorable to women, like secretariat jobs, caregiver jobs, and so forth are likely to be replaced by an intelligent machine, thus threatening the perceived jobs of women.

There is also the possibility of loneliness, social isolation, and mental degradation due to the upcoming social barriers posed by the Fourth Industrial Revolution. The possibility of a small technological world, occasioned by virtual and augmented reality, makes the physical world less interesting, posing a threat to the fabric of humanity. Recent online taxi businesses like Uber and other 100% digital-oriented organizations make social interaction in the workplace and the natural happiness that arises from making physical contact difficult. This could further lead to heightened stress, negative thoughts, and anxiety.

2.3.4 Environmental Risks

The environment is the biggest beneficiary of the positive and negative impacts of the Fourth Industrial Revolution development and technologies, whereby their raw materials come from the earth's natural resources and increased urbanization. According to PwC (2017), energy utilization, especially fossil-fuel energy from blockchain, autonomous vehicles, devices, sensors, and appliances, can devastate the environment, leading to climate change. Also, rural-urban migration resulting from the opportunities of the Fourth Industrial Revolution technologies will further worsen the state of rural areas, posing the risks of underdevelopment and degradation of existing development. The utilization of technologies resulting in rapid urbanization leads to the depletion of water, energy, and food resources, as well as other natural resources and GHG emissions. Using the technologies of the Fourth Industrial Revolution by companies will lead to industrial wastewater contamination, thus affecting the eight planetary boundaries of climate change. The eight planetary boundaries that would be affected are biosphere integrity, ocean acidification, ozone layer depletion, biogeochemical flows of nitrogen and phosphorus, atmospheric aerosol pollution, freshwater use, land system change, and the release of novel chemicals such as heavy metals, radioactive materials and plastics (Rockström et al. 2009). The World Economic Forum (2017) opined that human activities have caused "unprecedented environmental systems of change" over the years. The study noted current environmental risks of greenhouse gases, deforestation, ocean acidification, retreating of polar and glacier ice, pollution of nitrogen cycles, and water cycle pollution and abstraction. The activities of the Fourth Industrial Revolution will no doubt increase man's activities, thereby aggravating the already listed environmental risks and leading to environmental disequilibrium.

2.4 IMPACTS OF THE RISKS OF THE FOURTH INDUSTRIAL REVOLUTION

Risks always have impacts and consequences, which makes them risks in the first instance. The risks of the Fourth Industrial Revolution unequivocally have a ripple impact on people, industry, and the economy in such small dimensions, altering the world's status quo for all species and activities.

Industrially, the Fourth Industrial Revolution risks will change the existing business paradigm, activities, and functionalities. Organizations will face stiffer competition from organizations optimizing the technologies of the Fourth Industrial Revolution, especially through automation. This will lead to many firms closing down operations, downsizing, or being acquired in an unprofitable manner. Also, another impact is the loss of jobs, leading to an increase in the global unemployment rate and a worsened poverty rate in the world. Also, there will be a shrinking of stakeholders and a decrease in innovative ideas from stakeholders (employees, customers, suppliers, and other external environments).

The risks of the Fourth Industrial Revolution will also lead to high competition for raw materials, especially industrial espionage among companies and industry, especially high-tech companies, leading to scarcity and rising cost of raw materials. This will reduce the performance of organizations, the demand needs of the people, and the profitability of organizations, thus reducing the earning capacity of an organization's workforce. In addition, the transparency of the organizational process, transparency and accountability of financial flow, and the possibility of data storage may lead to the forceful takeover of companies, data compromise, susceptibility to fraud, and employee poaching. The Industrial Revolution led to more innovations and gave the technological capability to creativity. However, this will lead to the risks of over-customization of products, leading to value degradation of products and choice overload or buyers' remorse for potential and existing customers. Consumerism of this nature leads to product waste and waste of raw materials and its opportunity cost. 4IR will influence employees' procurement, utilization, and compensation, thus altering existing human resource management practices. However, this will lead to optimization but reduce the psychological perspective of human resource management in hiring and managing employees. Algorithms in the Fourth Industrial Revolution are binary and have limitations in evaluating the emotional perspective of human actions and activities. These will lead to poor judgment, low performance, reduction of team effectiveness, and low organizational performance.

The economic risks of the Fourth Industrial Revolution will lead to economic disruption, policy uncertainties, and ambiguity. Many economies, especially countries with low technological capability but high reliance on globalization, will have a tumultuous economy. World powers and the developed economy will use them as stabilizing indices for their economy. These will also lead to policy uncertainties and ambiguity, as the foundation of many 4IR-oriented economy policies will be based on guesswork or trial-and-error mechanisms. The Fourth Industrial Revolution can aid productivity by leapfrogging many economies from the traditional way of doing things. However, this comes at the cost of subserviency, as many economies will be at the mercy of high-tech companies and multinational organizations, controlling human activities and influencing the economic choices of humans. Many government policies will be based on the interests of world powers, high-tech companies, and organizations involved in utilizing large-scale 4IR technologies. The Fourth Industrial Revolution will bring about economic dislocation, whereby there will be investment disparity among different sectors of the economy, especially sectors such as energy, the built environment, transport, health, communication, agriculture, and forestry. These investment disparities will differ from country to country but share the same disadvantages of 4IR investment, not according to the needs but based on globalization correctness.

Economically, countries with poor transport infrastructure and inadequate logistics systems will be the losers during the disruption by 4IR on the supply chain. The supply chain activities will be disrupted, leading to the repackaging of spoilt/damaged products and the high cost of products. Many third-world countries are still grasping the integration of technologies into supply chain management, let alone the sophisticated and complex technologies of the 4IR. This change will further polarize the labor market and lead many people into poverty.

The risks of the Fourth Industrial Revolution will have a three-part impact on people, their relationship, sustainability concepts, and their health. The technologies will disfranchise people from one another, leading to social isolation or a pseudo-social augmented world. Many people will suffer from physical and mental depression as technology drives their relationships. Also, the disparity in blood relationships or friendship will escalate as technologies of the Fourth Industrial Revolution create new relationships for people across the globe. Many people will be too engrossed in the digital/cyber world than the physical world, whereby the latter becomes unattractive, even though the former is a pretense of the real world. This relationship dichotomy orchestrated by the 4IR will lead to the loss of family and societal values, as freedom from the digital world will threaten the freedom and rights of the people of the real world.

Moreover, people's relationships might be based on digital skills, as people with digital skills will be more closed than people without digital skills, which will also be felt in the workplace. Many people with digital skills will find it easy to communicate with people with digital skills. This can also lead to the proliferation of LGBTQs (lesbian, gay, bisexual, transgender, and queer), as same genders with digital skills will also relate to themselves, especially among men, as more men have digital skills than women.

Moreover, social and environmental sustainability will be impacted by the risks of the Fourth Industrial Revolution, as the technologies will threaten community cohesion regarding resource utilization and allocation. It will also adversely affect the ecosystem of man and the species around him, mostly water, energy, and food. Many chemical wastes and GHG emissions from high-tech companies currently affect and pollute the environment, which disrupts the ecosystem for man and the availability of resources, thus restraining the concept of sustainability in ensuring resources for future generation. Therefore, the Fourth Industrial Revolution will reduce environmental sustainability by increasing climate change, ozone layer depletion, water pollution, thermal pollution, increased deforestation, global warming, extinction of species in search of raw materials, biodiversity loss, coastal hazards, water scarcity, degradation of land, and desertification.

Furthermore, these environmental issues that the Fourth Industrial Revolution technologies can deteriorate people's health. People are exposed to harm, diseases, viruses, and sickness. These health concerns can lead to mass deaths of people, body disorders, eating disorders, and the weak hormone system of many people. The 2020–2021 coronavirus pandemic, because of biotechnology and genetic engineering, biological aspects of the 4th industrial revolution, led to millions of deaths. This shows that a single mistake from the technologies of the Fourth Industrial Revolution can lead to the extermination of a population or mass infection of a populace. This could also lead to biological war between highly oriented 4IR countries and underdeveloped countries. The population of countries will be at the mercy of the Fourth Industrial Revolution technologies.

2.5 SUMMARY

This chapter comprehensively introduced the technologies of the Fourth Industrial Revolution. The study vividly describes the Fourth Industrial Revolution and explains its main technologies. These technologies include the Internet of Things (IoT), artificial intelligence (AI), blockchain technology, cloud computing, robotic technology, and augmented reality. The chapter also espoused the risks of the Fourth Industrial Revolution. It categorized the risks into economic, human, social, and environmental risks. The chapter ended with the impacts of the risks of the Fourth Industrial Revolution, with a focus on its impact on the economy, industry, and people.

REFERENCES

Abdulhameed, O., Al-Ahmari, Ameen, W., and Mian, H. (2019). Additive Manufacturing: Challenges, Trends, and Applications. *Recent Trends in Design and Additive Manufacturing—Advances in Mechanical Engineering* 11(2), 1–27. https://doi.org/10.1177/1687814018822880.

Adepoju, O., Aigbavboa, C., Nwulu, N., and Olaiya, M. (2022). *Reskilling Human Resources for Construction 4.0. Implications for Industry, Academia, and Government*. Springer. https://doi.org/10.1007/978-3-030-85973-2.

Ahram, T., Sargolzaei, A., Sargolzaei, S., Daniels, J., and Amaba, B. (2017). Blockchain Technology Innovations. *2017 IEEE Technology & Engineering Management Conference (TEMSCON)*. IEEE. https://doi.org/10.1109/TEMSCON.2017.7998367

Bloem, J., Van Doorn, M., Duivestein, S., Excoffier, D., Maas, R., and Van Ommeren, E. (2014). *The Fourth Industrial Revolution: Things to Tighten the Link Between IT and OT Contents*. Atlanta, GA: Groningen Sogeti VINT.

Bottani, E., and Vignali, G. (2019). Augmented Reality Technology in the Manufacturing Industry: A Review of the Last Decade. *IISE Transactions* 51(3), 284–310. https://doi.org/10.1080/24725854.2018.1493244.

Brochal, F., Gonzalez, C., Komljenovic, D., Katina, P.F., and Sebastian, M. A. (2019). Emerging Risk Management in Industry 4.0: An Approach to Improve Organizational and Human Performance in the Complex Systems. *Complexity* 1–13. https://doi.org/10.1155/2019/2089763.

Carmigniani, J., and Furht, B. (2011). Augmented Reality: An Overview. In: Furht, B. (ed.) *Handbook of Augmented Reality*. Springer. https://doi.org/10.1007/978-1-4614-0064-61.

Chen, Y., Wang, Q., Chen, H., Song, X., Tang, H., and Tian, M. (2019). An Overview of Augmented Reality Technology. *IOP Conference Series* 1237, 1–5. https://doi.org/10.1088/1742–6596/1237/2/022082;www.researchgate.net/publication/334420829_An_overview_of_augmented_reality_technology.

David, L.O., Nwulu, N.I., Aigbavboa, C.O., and Adepoju, O.O. (2022). Integrating Fourth Industrial Revolution (4IR) Technologies into the Water, Energy & Food Nexus for Sustainable Security: A Bibliometric Analysis. *Journal of Cleaner Production* 363. https://doi.org/10.1016/j.jclepro.2022.132522.

Dechprom, S., and Jermsittiparsert, K. (2018). Foreign Aid, Foreign Direct Investment and Social Progress: A Cross Countries Analysis. *Opcion* 34(86), 2086–2097.

Dogo, E.M., Salami, A.F., Nwulu, N.I., and Aigbavboa, C.O. (2019). Blockchain and Internet of Things-Based Technologies for Intelligent Water Management System. In: Al-Turjman, F. (ed.) *Artificial Intelligence in IoT. Transactions on Computational Science and Computational Intelligence*. Cham: Springer. https://doi.org/10.1007/978-3-030-04110-6_7.

Fischhoff, B., Watson, S.R., and Hope, C. (1984). Defining Risk. *Policy Sciences* 17(2), 123–129.

Hegel, F., Muhl, C., Wrede, B., Hielscher-Fastabend, M., and Sagerer, G. (2009). Understanding Social Robots. *Proceedings of the 2nd International Conferences on Advances in Computer-Human Interactions*, ACHI 2009, IEEE, pp. 169–174.

Holotescu, C., Holotescu, V., and Holotescu, T. (2018). *Understanding Blockchain Technology and How to Get Involved*. Technical Report. https://doi.org/10.13140/RG.2.2.25185.33126/1.

Karabegovic, I., and Husak, E. (2018). Industry 4.0 Based on Industrial and Service Robots With Application in China. *Mobility and Vehicle* 44(4), 59–71.

Karjalainen, J., Heinonen, S., and Shaw, M. (2019). Peer-to-Peer and Circular Economy Principles in the Fourth Industrial Revolution (4IR)—New Risks and Opportunities. *2019 Proceedings of the 27th Domestic use of Energy Conference,* IEEE.

Koh, L., Orzes, G., and Jia, F.J. (2019). Guest Editorial. *International Journal of Operations & Production Management* 39(6), 817–828. https://doi.org/10.1108/IJOPM-08–2019–788.

Lee, M., Yun, J.J., Pyka, A., Won, D., Kodama, F., Schiuma, G., Park, H., Jeon, J., Park, K., Jung, K., Yan, M., Lee, S., and Zhao, X. (2018). How to Respond to the Fourth Industrial Revolution, or the Second Information Technology Revolution? Dynamic New Combinations Between Technology, Market, and Society Through Open Innovation.

Journal of Open Innovation: Technology, Market, and Complexity 4(21), 1–24. https://doi.org/10.3390/joitmc4030021.

Liao, Y., Loures, E.R., Deschamps, F., Brezinski, G., and Venancio, A. (2018). The Impact of the Fourth Industrial Revolution: A Cross—Country/Region Comparison. *Production* 28. https://doi.org/10.1590/0103–6513.20180061.

Lu, Y. (2017). Industry 4.0: A Survey on Technologies, Applications and Open Research Issues. *Journal of Industrial Information Integration* 6, 1–10.

McCarthy, J. (2007). *What Is Artificial Intelligence?* Stanford: Computer Science Department, Stanford University, pp. 1–15.

Min, J., Kim, Y., Lee, S., Jang, T., Kim, I., and Song, J. (2019). The Fourth Industrial Revolution and Its Impact on Occupational Health and Safety, Worker's Compensation and Labor Conditions. *Safety and Health at Work* 10, 400–408. https://doi.org/10.1016/j.shaw.2019.09.005.

Mitrovic, L.M. (2020). Challenges, Risks and Threats to Human Security in the 4th Industrial Revolution. *Journal of Criminalistics and Law, NBP* 25(1), 81–97. https://doi.org/10.5937/nabepo25-26316.

Mohammed, Z.K.A., and Ahmed, E.S.A. (2017). Internet of Things Applications, Challenges and Related Future Technologies. *World Scientific News* 62(2), 126–148.

National Academy of Science and Engineering—ACATECH. (2013). *Recommendations for Implementing the Strategic Initiative Industrie 4.0.* Final Report of the Industrie 4.0 Working Group. Frankfurt: ACATECH Report.

Onik, M.H., Kim, C., and Yang, J. (2019). Personal Data Privacy Challenges of the Fourth Industrial Revolution. *International Conference on Advanced Communications Technology (ICACT)*, February 17–19, pp. 635–638.

Outreville, J.F. (1998). The Meaning of Risk. In: *Theory and Practice of Insurance*. Boston, MA: Springer. https://doi.org/10.1007/978-1-4615-6187-3_1.

Oztemel, E., and Gursev, S. (2018). Literature Review of Industry 4.0 and Related Technologies. *Journal of Intelligent Manufacturing* 31, 127–182.

Patel, K.K., and Patel, S.M. (2016). Internet of Things—IoT: Definition, Characteristics, Architecture, Enabling Technologies, Application and Future Challenges. *International Journal of Engineering Science and Computing* 6(5), 6122–6131. https://doi.org/10.4010/2016.1482.

Poljak, D. (2018). Industry 4.0—New Challenges for Public Water Supply Organizations. *3rd International Scientific Conference*. Lean Spring Summit 2018, svibnja 24–25, 2018. https://www.bib.irb.hr:8443/1165868/download/1165868.Lean.Spring.Summit.Proceedings.2018.komplet.pdf#page=57.

Prisecaru, P. (2016). Challenges of the Fourth Industrial Revolution. *Knowledge Horizons Economics* 8(1), 57–62.

PwC. (2017). *Fourth Industrial Revolution for the Earth.* Harnessing the 4th Industrial Revolution for Sustainable Emerging Cities. PwC.

Rashid, A., and Chaturvedi, A. (2019). Cloud Computing Characteristics and Services: A Brief Review. *JCSE International Journal of Computer Sciences and Engineering* 7(2), 421–426. https://doi.org/10.26438/ijcse/v7i2.421426.

Renn, O., and Benighaus, C. (2013). Perception of Technological Risk: Insights From Research and Lessons for Risk Communication and Management. *Journal of Risk Research* 16(3–4), 293–313. http://dx.doi.org/10.1080/13669877.2012.729522.

Rockström, J., Steffen, W., Noone, K., Persson, Å., Chapin III, F.S., Lambin, E., Lenton, T.M., Scheffer, M., Folke, C., Schellnhuber, H.J., and Nykvist, B. (2009). Planetary Boundaries: Exploring the Safe Operating Space for Humanity. *Ecology and Society* 14, 32. https://doi.org/10.5751/ES-03180-140232.

Ruzarovsky, R., Holubek, R., Janicek, M., Velisek, K., and Tirian. (2020). Analysis of the Industry 4.0 Key Elements and Technologies Implementation in the Festo Didactic Educational Systems, MPS 203 14.0. *Journal of Physics, International Conference on Applied Sciences (ICAS 2020)* 1781. https://doi.org/10.1088/1742–6596/1781/1/012030.

Schwab, K. (2016). How Will the Fourth Industrial Revolution Affect International Security? www.weforum.org/agenda/2016/02/how-will-the-fourth-industrial-revolution-affect-international-security/. Accessed 20 September 2021.

Silva, R., Oliveira, J.C., and Giraldi, G.A. (2003). Introduction to Augmented Reality. www.researchgate.net/publication/277287908_Introduction_to_augmented_reality. Accessed 19 October 2021.

Soh, C., and Connolly, D. (2021). New Frontiers of Profit and Risk: The Fourth Industrial Revolution's Impact on Business and Human Rights. *New Political Economy* 26(1), 168–185. https://doi.org/10.1080/13563467.2020.1723514.

Srinivas, J., Reddy, K.V.S., and Qyser, A.M. (2012). Cloud Computing Basics. *International Journal of Advanced Research in Computer and Communication Engineering* 1(5).

Stancioiu, A. (2017). *The Fourth Industrial Revolution—Industry 4.0*. Fiabilitate si Durabilitate—Fiability & Durability No. 1, pp. 74–78. www.utgjiu.ro/rev_mec/mecanica/pdf/2017-01/11_Alin%20ST%C4%82NCIOIU%20-%20THE%20FOURTH%20INDUSTRIAL%20REVOLUTION%20%E2%80%9EINDUSTRY%204.0%E2%80%9D.pdf.

TutorialsPoint. (2015). *Artificial Intelligence, Intelligent Systems*. Tutorials Point (I) Pvt. Ltd. www.tutorialspoint.com.

Umachandran, K. (2020). Industry 4.0 and Its Associated Technologies. *Journal of Emerging Technologies (JET)* 1(1), 1–10.

UNDP. (1994). *Human Development Report 1994*. Oxford: Oxford University Press.

Wild, J., Arnold, M., and Stafford, P. (2015). Technology: Banks Seek the Key to Blockchain. *Financial Times*. https://www.ft.com/content/eb1f8256-7b4b-11e5-a1fe-567b37f80b64.

Williams, B. (2021). *An Introduction to Robotics*. Dr. Bob Productions. Athens, OH: Ohio University, pp. 1–46.

World Economic Forum (WEF). (2017). Harnessing the Fourth Industrial Revolution for the Earth. www.weforum.org/projects/fourth-industrial-revolution-and-environment-the-stanford-dialogues. Accessed 10 October 2021.

Xu, M., David, J.M., and Kim, S.H. (2018). The Fourth Industrial Revolution: Opportunities and Challenges. *International Journal of Financial Research* 9(2), 90–95. https://doi.org/10.5430/ijfr.v9n2p90

Yaga, D., Mell, P., Roby, N., and Scarfone, K. (2018). *Blockchain Technology Overview*. National Institute of Standards and Technology. https://doi.org/10.6028/NIST.IR.8202.

Zervoudi, E.K. (2021). Fourth Industrial Revolution: Opportunities, Challenges, and Proposed Policies. In: *Industrial Robotics—New Paradigms*. IntechOpen, pp. 1–25. http://dx.doi.org/10.5772/intechopen.90412.

Zheng, Z., Xie, S., Dai, H., Chen, X., and Wang, H. (2017). An Overview of Blockchain Technology: Architecture, Consensus, and Future Trends. *2017 IEEE 6th International Congress on Big Data*, pp. 557–564. https://doi.org/10.1109/BigDataCongress.2017.85.

3 Introduction to Risk Management and Current Risk Management Framework

3.1 INTRODUCTION TO RISK MANAGEMENT

Risk and risk management are unavoidable aspects of human living and project endeavors. Risks can create a positive or negative impact on projects and organizations, with its history dating as far back as after the Second World War with market policies to safeguard persons and businesses from several losses associated with accidents using market insurance. Risk and risk management became a scientific discipline in 1950 when most business owners realized that it was expensive and almost impossible to manage all risks with insurance. This led to the development of several alternatives, such as contingent planning and self-protection activities, coverage for work-related illnesses and accidents, and training and safety programs to complement insurance and reduce risk exposure. This process evolved into identifying several forms of risk, such as operational, financial, economic, technological, environmental, brand, legal, reputational, liability and property losses, credit, and market risk, among others. Concurrently, administering risk management became important, leading to integrated risk management. Risk management has undergone significant development to become a responsive and self-regulating feature necessary for organization development and strategic decision-making.

Risk has become a vital concept in various fields of operation, and it has several definitions. Srinivas (2019), among several meanings, as shown in Table 3.1, defined risk as uncertain events that may positively or negatively impact project objectives. Risks are situations or conditions in which the occurrence in all rational foresight will lead to a contrary effect or impact on any aspect of project implementation. The International Organization for Standardization (ISO) defines risk as the effect of uncertainty on objectives (Risk Management, ISO 2009). Risk also refers to the uncertainty and severity of an activity's actions and outcomes. It can also be defined as an uncertain event or set of events that can influence the achievement of objectives when it occurs, and it is commonly expressed as a product of the probability of the occurrence of an adverse event and the weight of the consequences of that event (Sotic and Rajic 2015).

DOI: 10.1201/9781003522102-4

TABLE 3.1
Different definitions of risks

Serial Number	Source	Definition
1	Project Management Institute	An unpredictable occurrence or circumstance that, if it materializes, could positively or negatively impact the project's goals.
2	Institute of Risk Management	A combination of an event's likelihood and its effects.
3	Association of Project Management Body of Knowledge	Project risk is a hazard that, if it materializes, could have either a favorable or unfavorable impact on the project's goals. A risk has a source and, if it materializes, a result.
4	British Standard BS IEC 62198:2001	Combining the likelihood that an event will occur and how it would affect the project's goals.
5	Business Dictionary	A likelihood or potential for harm, injury, liability, loss, or any other unfavorable event that is brought on by external or internal vulnerabilities and is preventable by preventative action.
6	Fundamentals of Risk Management	Risk is defined as "being exposed to danger," and the definition of risk is "a potential or possibility of danger, loss, harm, or other unfavorable effects." However, taking a chance might also have a good effect. The risk associated with outcome ambiguity is a third possibility.
7	Adams	Risk is "the likelihood that a specific unfavorable occurrence occurs during a specified period or results from a specific challenge."
8	Philosophy of Risk	Risk has been interpreted as Risk = Hazard + Exposure, where Hazard is defined as how a thing or circumstance can cause harm. Exposure is the degree to which the hazard can affect the likely recipient of the harm.

Source: Srinivas (2019)

Aven (2016) described the highlights of risk definition, including the probability of an unsuccessful event, the likelihood of undesirable, adverse consequences of an event, the occurrence of a loss of which one is uncertain, the consequences of an action and its accompanying uncertainties, doubt about and severity of the consequences of an activity concerning something that humans value, the occurrences of some specified consequences of the activity and associated uncertainties, and the nonconformity from a reference value and related uncertainties. Risk can be categorized as preventable risk that emanates inside the organization. These risks arise from workers' or manager's unethical, unlawful, or inappropriate behaviors or actions and failures in routine operational procedures. This type of risk can be managed, eliminated, or avoided. Another risk classification is strategic risks that companies undertake of their own accord to generate higher yields from investment. Strategic risks are sometimes referred to as opportunity cost risks. Unlike preventable risk, strategic risk is undesirable but unavoidable for organizational growth. For example, a strategy with high projected proceeds will usually require the company to accept some significant risks, and managing such risks is a determinant in securing the intended gains. External risks, also known as absolute risks, on the other hand, arise from events outside the company that are beyond the company's influence or control. External risks emanate from natural disasters, government policies, or major natural, economic, political, or social changes. Risks are generally unavoidable and can be fatal to a company's strategy and even to its survival if not properly managed, hence the development of risk management to minimize or mitigate risks.

Risk management can be defined as a planned and organized practice intended to help the project team make accurate decisions at the appropriate time to recognize, categorize, quantify the risks and develop measures to manage and control them (Srinivas 2019). Risk management is a critical component of management and accountability, and it involves identifying, evaluating, and prioritizing risks, followed by the coordinated and rational use of resources to reduce, manage, and control the probability and the consequences of untimely events. The concept of risk management denotes the view that the probability of an occurrence can be reduced and the impacts curtailed. Effective risk management endeavors to assuage the risk while capitalizing on the proceeds of the risk. Risk management encompasses pinpointing risks, evaluating their consequences, arriving at the best course of action, and assessing the risk (Owojori et al. 2011; Zidafamor 2016). It is simply the process of determining or measuring risk and developing policies and plans to manage the risk. Risk management may also be defined as identifying, examining, and prioritizing risks tracked by harmonized and cost-effective use of resources to oversee and control the possibility and outcome of ill-fated happenings.

A risk management plan involves developing measures to recognize, evaluate, and control threats to an organization's capital and earnings. These risks or threats can arise from various causes, such as strategic management inaccuracies, legal liabilities, economic uncertainty, accidents, and natural disasters. Petr and Blanka (2018) defined risk management as a logical or systemic approach to identifying, evaluating, assessing, ranking, and monitoring risks through a project lifecycle. It involves a systematized and inclusive approach focused on identifying, organizing, and responding to risk factors to achieve project goals; hence, it is considered a proactive rather than a reactive concept. Over the years, the concept of risk management has become widely accepted in several companies. Many organizations have regularly affected risk management in projects to enhance productivity, profits, and increase business performance. Risk management implementation helps organizations deal with imminent concerns, fears, and uncertainties. It is important to know that the concept of risk management cannot be isolated from the concept of planning and prioritizing. The risk management process encompasses the methodical application of managerial guidelines, procedures, and processes for appropriate responsibilities, identification, investigation, appraisal, conduct, monitoring, and risk communication (Bahamid and Doh 2017). Effective risk management helps an organization improve relationships between employees, clients, team members, contractors, and society, improve employees' confidence and chances of success in a work environment, reduce organizations' or project's exposure to liability, and understand and manage project requirements and outcomes.

However, this book is geared toward managing technological risks, which are risks incurred while utilizing technology to accomplish stated project goals. This is highly necessary due to the risks' devastating effects and undesirable consequences. Nevertheless, the approach to managing risks is similar in that the technological risks of the Fourth Industrial Revolution are not exempted.

Consequently, the goal of risk management is not to remove all project risks altogether. It aims to produce an organized framework that will make management manage project risks more efficiently and effectively, most notably the crucial ones. Risk management is the merger of approaches, methodologies, and processes to identify, evaluate, and select alternative supervisory and practical responses to risk. The risk management process, however, does not eliminate all project risks, but it is focused on producing a systemic framework to manage risks, most essentially the critical ones, in a more organized and active way, thereby increasing the chances of attaining success in project endeavors. The risk management process aims to ensure the viability of the project before inception and the project into the future to predict and analyze "worst-case" scenarios (Al Ariss and Guo 2016). The risk management process supports companies in reducing unforeseen risks and setting up standardized project implementation procedures. Research has revealed that risk management impacts project success, organizational performance, scheduled time, project budget, and project requirements and specifications compliance.

The traditional risk management process developed earlier hinged on protecting companies from loss through conformance procedures and hedging techniques while avoiding the negativities that come with risk. However, the new risk management process model is focused on the gains and opportunities associated with risk while managing the negativities. The risk management process comprises five stages, as shown in Figure 3.1:

1. Risk identification/classification
2. Risk analysis
3. Risk response
4. Risk control
5. Risk review

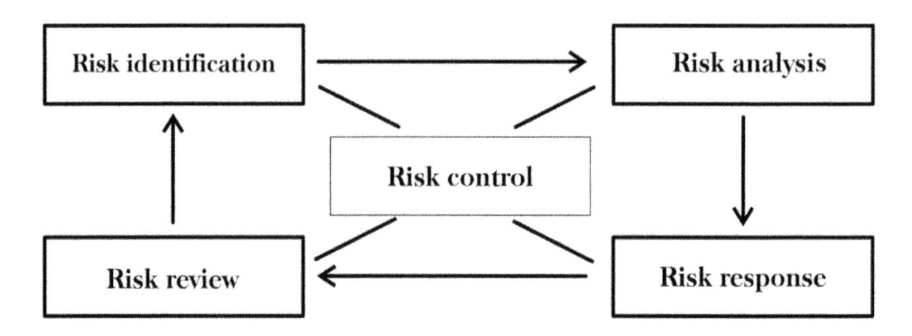

FIGURE 3.1 The risk management process (Alsaadi and Norhayatizakuan 2021, p. 5)

3.2 RISK IDENTIFICATION PROCESS

Risk identification is the process of recognizing risks, which includes detecting the sources of risks, pinpointing the risk events, and ascertaining the impact of the risks related to attaining project goals (Djamaluddin et al. 2020). The process of risk identification is the core of developing an appropriate risk management plan; hence, it requires the significant contribution of all stakeholders. However, in the case of a technological risk, these stakeholders might be internal stakeholder's firsts, such as the team in charge of the technology, and external stakeholders, such as the beneficiaries of the technology. It is mostly from internal stakeholders such as product owners, business analysts, software developers (both frontend and backend), UX/UI designers, scrum masters, and other executive members. Liu et al. (2016) stated that risk identification is the process of analyzing and constantly identifying, assessing, and categorizing the initial importance of the risks related to construction projects and the interrelationships among these risks. The process of risk identification involves identifying and recording information associated with risks and prioritizing potential risks according to the impending impact on the project or organization's objective. The process of identifying risk is the most crucial step in the risk management process. It is worthy to note that risk identification is not used to make perfect predictions about future events or occurrences; rather, the purpose is to highlight probable causes of risk that would greatly impact the project or organization's objective irrespective of the technology deployed. Consequently, identifying risk ensures that potential risks are properly managed to achieve the overall project or organizational objectives (Dario 2017).

The process of risk analysis and response management process is built on successfully identifying and classifying potential risks. Failure in the risk identification stage can result in dire consequences on project development and its success which can lead to critical impacts on the resources available to the organization. The process of risk identification helps organizations or projects to identify risks and the impacts they can have on projects or organizations, identify the most important input data, possess enhanced information on the relevance of the process, and provide adequate information for decision-making (Rostami 2016). The process of risk identification is mostly subject to experience or lessons learned from past projects and study of related projects. According to George (2020), an effective risk identification process requires that stakeholders widen their information sources to capture enough risks that might pose a threat to the project and be conversant in detecting risks, as this will help the team to know the areas to pay attention to during the risk identification. Stakeholders should be willing to explore all available risk identification techniques and tools to ensure a robust identification process and that all identified risks are adequately documented to ensure that the risk identification process is effective and efficient. Risk identification and classification can be achieved using documentation reviews or information-gathering techniques, which this book recommends in identifying any risks within an organization or during item generation from another technological product. Documentation review involves a structured appraisal of project documents, historical study of similar projects execution and quality of plans as well as the uniformity between project plans and project requirements/assumptions, while information gathering techniques include brainstorming, Delphi technique, cause-and-effect analysis, questionnaires, scenario analysis, checklist analysis, direct observations and surveys, structured what-if technique (SWIFT), expert judgment, incident analysis, bow tie analysis, SWOT analysis, interviews, fault tree analysis (FTA), presumption analysis, and so forth.

1. **Brainstorming:** One of the most well-known risk identification methods is brainstorming. PMI (2013) states that this technique is typically carried out by a multidisciplinary group of experts outside the team, and getting a thorough inventory of project risks is the aim. Here, everyone pertinent to the project is gathered in one location. There is only one facilitator, and she or he briefs the participants on several components before noting the factors (George 2020). The two stages of brainstorming are as follows: (i) the idea generation phase, during which participants come up with as many concepts as they can; (ii) the concept selection phase, where each member argues for their chosen idea to persuade the others. The ideas are filtered in this second round, leaving only those that have received unanimous group approval. Four fundamental guidelines govern this method: (i) Criticism is prohibited; evaluation of ideas must be postponed; (ii) "freewheeling" is encouraged; (iii) quantity is desired; the more ideas there are, the higher the likelihood that some of them will be helpful; (iv) combination and improvement. This technique involves an open conversation attended by project teams and other project participants, giving the impression that it is a well-organized risk detection technique. As a result, it offers a chance to discuss the risks' presence and potential effects. However, if it is not being watched, it is prone to be affected by more powerful parties (Garrido et al. 2011).
2. **Interviews/expert opinions:** This entails unstructured, semi-structured, or structured interviews performed individually or in groups with knowledgeable project participants, experts, beneficiaries, and/or stakeholders.

This in-person interview can assist in identifying several project risks that are hidden from view. Consulting experts or speaking with experienced project participants can be quite helpful in avoiding or resolving recurring issues.

3. **Delphi technique:** Delphi is a method for getting experts' opinions on future occurrences in agreement. A well-organized collective judgment is superior to an individual opinion, and this claim is supported by structured knowledge, experience, and creativity from an expert panel. This idea is based on an estimation method in which a predetermined stop condition allows a structured group of subject experts to respond to questions in two or more rounds. Each panelist will get an anonymous question from the organizer after each round, who will then decide or judge the answer. The range of alternatives does, however, keep narrowing during this process until the group comes to a consensus on a specific response, which is then accepted as the right one. The physical group gathering is replaced with written responses in this consensus-building method. This method calls for the systematic collection and critical comparison of judgments about a particular issue from geographically isolated anonymous participants using a series of precisely crafted surveys interspersed with summary data and feedback drawn from prior responses. This approach might be helpful in identifying risks, but it works better for tying the likelihood of occurrence and potential consequences to already known risk events (Jayasudha and Vidivelli 2014).
4. **Questionnaires:** A project's potential dangers can be found using the risk identification questionnaire technique. Project team members receive attribute-level questions, particular hints, illustrations, and inquiries for further research. The questionnaire is often designed to be specific to each software development project and each development stage. There are two steps to applying for the questionnaire: Phases 1 and 2 are for questions and answers and for issuing clarification. It permits uniformity, rapid response times, and candid exposure of hazards. The main drawback is that the outcomes depend on people's opinions (Garrido et al. 2011; Renault et al. 2016).
5. **SWOT analysis:** SWOT analysis is a planning method to assess a company's strengths, weaknesses, opportunities, and threats. It categorizes internal and external factors dependent on achieving project objectives. Opportunities and threats are assessed externally, while the organization's internal strengths and weaknesses are considered. Risks are the challenges and opportunities that the project faces (Renault et al. 2016).
6. **Checklist analysis:** It consists of a list of statements marked "yes" or "no," which can be utilized by an individual project team member, a group, or during an interview. The checklist can be created using historical data, expertise gained from prior initiatives with similar scope, and information from other sources (Gajewska and Ropel 2011).

7. **Documentation review:** This requires periodic evaluations of past projects, considering all the assumptions, plans, and prior project files. These could be used as markers to show deeply ingrained project dangers (PMI 2013).
8. **Scenario analysis:** Scenario analysis is the creation of fictitious scenarios that illustrate the processes that would emerge from the logical formulation of each event, as well as its interactions and outcomes, which distinguishes it from other approaches. It is essential to determine whether risk events can happen simultaneously and whether there are significant differences between them; find the risk trigger, which is the factor that makes a combination of factors produce high and low risks; create a scenario containing uncertainty variables, correlate them, and determine how it will affect the project; Identify the risk elements, such as new technology, potential virus, cyber-attack nodes, overly optimistic estimating, or a potential workforce issue; calculate how risk triggers will affect the project's goals; and incorporate the likelihood of potential events and the correlation between them using simulation techniques (Garrido et al. 2011).
9. **Cause-and-effect diagrams:** This diagram shown in Figure 3.2, sometimes referred to as Ishikawa diagrams or fishbone diagrams, shows how different components may be connected to potential issues or impacts

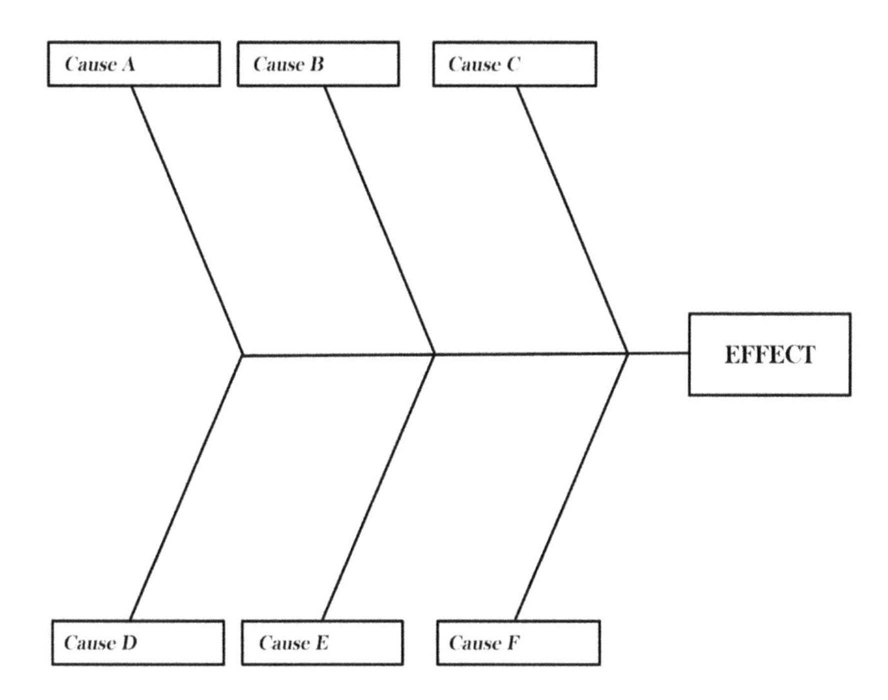

FIGURE 3.2 Cause and effect/fish bone diagram (Coccia 2018, p. 5)

(PMBOK–PMI 2008; Coccia 2018). The diagram is constructed by listing the causes on the left and the effects on the right. The primary causes must be categorized in accordance with the categories for each impact.

10. **SWIFT analysis:** The structured what-if technique (SWIFT) is a group-based, methodical investigation where a facilitator uses a series of terms or brief sentences inside the workplace to encourage collaboration in identifying hazards. The original SWIFT was created as a more user-friendly alternative to the HAZOP (hazard and operability) technique, which is used to identify risks and issues relating to the operability of installations. Recent development in SWIFT typically allows for its use at several system levels and requires less data than HAZOP. The SWIFT technique requires the facilitator and the group to employ common expressions like "what-if" together with single words or short phrases. This is done to determine how abnormal behavior or regular operation affects the system, industry, organization, or proceedings.
    (IEC/ISO 31010 2009; Garrido et al. 2011)

Using each of these techniques to identify potential risks that could impact the project's goals is possible. Since new risks may emerge as the project moves through its lifespan, risk identification is iterative. The prior dangers would have likely been handled as these new threats emerged. The efficient management of project risks requires the participation of all parties.

3.3 RISK RESPONSE PLANNING AND TECHNIQUES

The process of creating actions and measures to improve opportunities and lessen threats to project objectives is known as risk response planning. The main advantage of this method is that risks are dealt with according to priority, with resources and activities added to the budget, schedule, and project management plan as necessary (Rehacek 2017; David et al. 2023). Risk response is a crucial element in the risk management process, and it is impossible to manage risks in investment projects efficiently without having a working knowledge of how to deal with them. Finding an adequate risk response requires a diverse strategy to account for the potential negative effects of different solutions. Approaches that are commonly used to address threats or risks, if they materialize, that could have a detrimental influence on the project's objectives include the following:

1. Risks avoidance or elimination
2. Risk transfer
3. Risk mitigation or reduction
4. Risk acceptance

Risks avoidance or elimination: Risk avoidance can be defined as the elimination of activities with a high likelihood of loss by making it difficult for risk to occur or by carrying out the project in a different way that will achieve the same goals but shield the project from the influence of the risk. Risk is avoided by establishing effective control mechanisms that replace risk-bearing actions with other less risky or non-risky behaviors that are less risky or non-existent while accomplishing the desired goals. This process excludes the chance for risk and any negative effects and prospective benefits that might be gained concurrently with the risk.

Risk transfer: This entails transferring the risk to a third party, who will manage it and its effects. It merely shifts the liability to a different party, not reducing the risk. This can be accomplished by purchasing insurance, in which case the insurance provider is now liable, or having the work done under a fixed-price contract, where the contractor will be held liable. This method creates the issue of appropriate analysis of the established actions since risk transfer might result in other hazards that need to be examined and monitored.

Risk mitigation or reduction: Risk mitigation is the process of lowering risk by lowering its impact to make it more tolerable to the project or organization. It requires actions taken to lessen the risk's impact and likelihood and the steps required to repair any damage the risk can cause in a controlled fashion by developing an intervention plan with concrete mitigation measures. Measures for preventing fraud and corruption, compensatory damages, and retroactive or prospective payment of losses can all be incorporated here.

Risk acceptance: Risk acceptance occurs when it is clear that there is no way to avoid the risk, when the costs associated with risk removal are too huge for the organization or project to bear, or when using an alternate response plan is impossible. The framework for the risk management process is built on the analysis and evaluation of risks utilizing one or more of the acceptance principles of using best practice codes, comparison to related reference systems, and explicit estimation of the risk. All these factors indicate that the management must determine how far the company can go regarding risk acceptance from the beginning of developing risk strategies. Risk acceptance is done by either actively allocating suitable contingency or inaction while only keeping track of the risk.

These risk response techniques each have a different and distinctive impact on the risk condition. Risk acceptance can be applied to both positive and negative risks and possibilities these tactics are adopted in accordance with the likelihood of the risks and their implications for the project's overarching goals. Transference and acceptance are typically appropriate strategies for dangers that are less critical and have a low overall impact. At the same time, avoidance and mitigation techniques are good strategies for critical risks with significant impact (Ahmadi-Javid et al. 2020).

Several project documents are updated as necessary during the plan risk responses process. For instance, relevant risk responses identified and approved are added to the risk register. The planning process's deliverable becomes the risk register. It is a tool used by the project team to record risks and offer appropriate addressing methods to lessen their impact should they materialize. The risk register covers risk categories, risks that have been detected, risks that have caused problems, and risks that have been addressed, providing a thorough picture of how to manage risk in projects. The high and intermediate dangers are frequently covered in great detail. Risks that are considered low priority are added to a watch list for recurring monitoring. The risk register should be written in full according to the priority ranking and the anticipated response (Piotr 2012).

3.3.1 Importance of Risk Register

1. Facilitates smooth communication among project team members regarding project goals, costs, risks, and so forth.
2. Helps project stakeholders better comprehend the effects of project hazards detected.
3. As the project moves forward, being able to track each risk's status allows the team to better monitor and manage risks.
4. Helps management analyze and update project risks and serves as a resource for upcoming initiatives.
5. Assists the project team in deciding how to handle each risk in light of its seriousness.
6. Contributes strategically to project decision-making. Management will determine whether to move on with the project based on the types and characteristics of risks outlined in this document.
7. Provides guidance to the project team regarding risk management throughout the project.

3.4 QUALITATIVE AND QUANTITATIVE RISK ANALYSIS

Risk analysis involves analyzing how project outcomes and objectives might change because of the effect of the risk occurrence. After risks are identified, they are analyzed to determine their qualitative and quantitative effects on the project so that the proper steps can be taken to reduce them. It investigates the ambiguity of prospective hazards and how, if they materialized, they would affect the project's schedule, quality, and costs. It's crucial to keep track of hazards over the course of a project because risk analysis is not a precise science. Generally, risk can be analyzed in two different ways—quantitatively and qualitatively—taking consideration of the statistics and managerial aspects of technological risks.

3.4.1 Qualitative Risk Analysis

In qualitative risk analysis, risk factors are ranked from highest to lowest based on their likelihood of happening and impact. The definitions from the risk management plan are used in this procedure, and raw types of impacts and probability are manually classified risks and not by computed values with respect to the entire project and the side effects of other risks. Qualitative risk analysis establishes hazards relative importance for future investigation or action by evaluating and combining risk impact and probability of occurrence. Project managers can focus on high-priority risks and lower the amount of uncertainty thanks to this procedure, which is its main advantage (Evrin 2021).

In ensuring successful risk management, qualitative risk analysis depends on several variables such as awareness of the risks, dangers, and issues associated, as well as the persons involved, the frequency of exposure, the exposure levels, the competence of the individuals, prior statistical information, and any current control mechanisms (IOSH 2018). Utilizing a qualitative approach, risk areas pertaining to standard business operations can be rapidly identified. The method can occasionally be challenging due to the staff members' prejudices or their lack of work experience, but qualitative risk analysis typically increases an efficient risk assessment approach. Qualitative risk assessment can be quickly implemented because there are no measures or statistical or numerical dependencies. Tools used for qualitative analysis include risk probability and impact assessment, probability and impact matrix, risk data quality assessment, risk categorization, risk urgency assessment, and expert judgment. However, in utilizing these tools, the following techniques are necessary.

3.4.1.1 Bow Tie Analysis

This qualitative risk analysis technique identifies all potential project hazards and their sources and effects. The project management team must first identify risks that could impact the project before considering their sources, effects, and—more crucially—a risk mitigation approach. It is an extremely adaptable technique that may be applied in any sector.

3.4.1.2 Risk Register

In managing technological projects or projects that will utilize any of the Fourth Industrial Revolution technologies, a risks register is a project management tool needed, which records risks associated with the project. In this document, all potential hazards that can arise during the project execution phase are listed together with essential details. It is intended to be utilized as a component of the risk management plan, which outlines who oversees managing the risks, how to mitigate them, and what resources are required. Several trustworthy information sources, including the project team, subject matter experts, and historical data, are typically used to create a risk register.

3.4.1.3 Risk Analysis Matrix

The risk analysis matrix depicted in Figure 3.3 assesses the likelihood and the severity of identified risks, classifying them by order of importance. Its main purpose is to help managers prioritize risks and create a risk management plan with the right resources and strategies to properly mitigate them. Risk likelihood is measured on a relative scale, not a statistical one, which makes it a qualitative risk analysis tool.

		Consequences→				
		A	B	C	D	E
Likelihood→	V	Medium 5	High 10	High 15	Extreme 20	Extreme 25
	IV	Medium 4	Medium 8	High 12	High 16	Extreme 20
	III	Low 3	Medium 6	Medium 9	High 12	Extreme 15
	II	Low 2	Low 4	Medium 6	High 8	Extreme 10
	I	Negligible 1	Low 2	Medium 3	High 4	Extreme 5
Consequence Scale 1		Description 1	Description 2	Description 3	Description 4	Description 5

FIGURE 3.3 Risk matrix (Peace 2017, p. 4)

3.4.1.4 Benefits of Qualitative Risk Analysis

1. The assessment team involves people directly impacted.
2. Countermeasure strategies can be implemented more quickly.
3. Subject matter specialists can aid in a better understanding of the dangers.
4. The risk management strategy does not depend on or prioritize direct costs.
5. Risk analysis typically takes place at the source.
6. Builds a strong team culture using human expertise and experience.

3.4.1.5 Limitations of Qualitative Risk Analysis

1. Risks are rated subjectively.
2. The subjective evaluation is based on the participants' abilities.
3. There may be variations in each person's interpretation.
4. It involves evaluating a limited sample size, which might not be representative of all events.
5. It is challenging to assess any cost-benefit ratios.
6. The likelihood and impact scores contain a lot of assumptions.
7. Assumptions might be inaccurate due to limited experts.

3.4.2 Quantitative Risk Analysis

Quantitative risk analysis involves calculating the impact of recognized risks on the overall project goals. This method's main advantage is that it generates quantifiable risk data, primarily numerical, to aid decision-making and lessen project uncertainty. The quantitative method uses data to fit objective probability distributions using a set of parameters to a model. It is a method of numerically examining the effect of recognized risks on overall project objectivities. The model's behavior is then examined using a probability simulation tool, which uses data and parameters to generate a collection of simulated outcomes (Garnett 2018).

Quantitative risk assessment is based on realistic and measurable facts used to compute the impact values that the risk will create with the chance of occurrence; it offers more objective information and accurate data than qualitative analysis (Evrin 2021). Risks that have been prioritized are subject to quantitative risk analysis, which examines the consequences of certain risks and events and rates each risk numerically. In the quantitative risk analysis process, impacts on the entire project will be computed to get a more developed total ranking instead of evaluating the individual impacts using raw typology. Techniques like the simple multi attributes rating technique (SMART) or decision-making theory are helpful (PMBOK 2013).

Tools used for quantitative analysis include the following:

1. The tornado diagram is used to illustrate the sensitivity analysis results.
2. Expected monetary value analysis is a statistical approach that determines the average result when the future contains situations that may or may not occur.
3. Decision tree analysis is typically constructed using a decision tree diagram that depicts a situation under evaluation and the ramifications of each available choice and potential scenario.
4. Modeling and simulation to calculate the full impacts. For instance, using the Monte Carlo analysis.

3.4.2.1 Benefits of Quantitative Risk Analysis

1. It provides a means of examining the cumulative impact of risks on objectives instead of evaluating each risk separately and individually.
2. It describes risks statistically by using numbers or ranges for probability and impacts rather than imprecise phrases like "high" or "low."
3. It offers consistency in the analysis since the simulation model's operation and results are independent of the person conducting the analysis and are not influenced by biases and subjective beliefs.
4. Allowing the model to be flexible and examine various situations and alternatives to the base case enables the investigation of multiple solutions for risk management.
5. It allows the creation of a complicated model of reality capable of making precise predictions by reflecting a level of complexity that is more than what a single human can comprehend or remember.
6. Setting realistic goals and presenting them as a spectrum of potential outcomes rather than a single point.

3.4.2.2 Limitation of Quantitative Risk Analysis

1. The requirement for software tools adds to the project's costs and probably staff training to utilize them well and requires integration with other project tools.

2. Analytical outputs require careful interpretation, which may require some knowledge of statistical principles.
3. Since computer-based technologies inherently produce outputs with many decimal places, the results have a falsely high level of precision that is unlikely to be supported by the input data.
4. Giving excessive weight to model outputs without exercising enough critical thought or judgment on the outcomes.
5. Using specialist technologies may lead to relying on an expert to run the analysis and carry out duties that the team does not fully comprehend, creating a sense of detachment and a loss of ownership.

3.5 THE PESTLE ANALYSIS

The PESTLE analysis is an approach for identifying and analyzing the significant forces causing a change in the organizational environment. It is a form of risk appraisal that aids in identifying environmental risks to a technological project or any project that requires integrating and utilizing the technologies of the Fourth Industrial Revolution. This is further discussed in Chapter 10. The PESTLE analysis is an acronym for political, economic, social, technological, legal, and natural environment. PESTLE analysis is an organizational audit of the company's operations to identify the various forces and elements in the outside environment that influence an organization's success. The tool enables evaluation of the current environment and anticipated changes. It is founded on the idea that businesses that constantly scan their environment get an advantage over rivals since doing so makes it easier to gather, process, and use the information to boost productivity. More specifically, it is employed in forecasting and assessing the future condition of affairs. Since it looks at the external environment of the organization's opportunities and threats, it is typically seen as a component of the SWOT analysis process. It is a component of the external assessment, appraisal, review, and evaluation, as well as the interests of stakeholders with a stake in politics, economics, social issues, law, and business that can be recognized through the analysis and are reflected in the findings (Buye 2021; Ronald 2021).

It is especially helpful when a company starts or presents a new product. PESTLE analysis can help a company become familiar with the opportunities presented by the current circumstances in the organizational environment. It can also be used to pinpoint present or potential problems, allowing for efficient planning on handling them. The analysis' findings can help areas where improvements or modifications are needed.

1. **Political factors:** This entails the degree to which a government may affect the economy or a particular business decision regarding technology deployment, especially in terms of policies. The entire revenue-generating structures of organizations may alter because of a government imposing a new tax or duty, for instance. Political considerations

include tax laws, fiscal policies, trade tariffs, and other measures that a government may enact throughout the fiscal year, which may significantly impact the business environment.

2. **Economic factors:** These economic variables directly affect a company's success and have long-lasting economic reverberations. For instance, a rise in any economy's inflation rate would impact how much businesses charged for their goods and services. Additionally, it would impact consumer purchasing power and alter that economy's demand and supply patterns. Inflation, interest rates, foreign exchange rates, economic development patterns, and so forth are all examples of economic factors. It also considers foreign direct investment, depending on the industries being examined.
3. **Social factors:** The sociological element considers all occurrences that have a social impact on the market and community. As a result, it is also necessary to consider the project's benefits and drawbacks for the local populace. These occurrences cover things like global warming, population dynamics, healthy consciousness, and cultural expectations and standards. These variables examine the market's social environment and evaluate influences, including societal trends, demographics, population analytics, and so forth.
4. **Technological factors:** These elements relate to technological advancements that could positively or negatively impact the market and industry. This refers to technology knowledge levels within a market, automation, and research and development. This component considers all technologically relevant occurrences. It is crucial to consider this because technology frequently becomes outdated a few months after it is released, as most risks might be generated from associated technological solutions. This component might also consider any entrance obstacles to certain markets and adjustments to financial choices.
5. **Legal factors:** This component considers all aspects of the law, including employment, quotas, taxes, resources, imports, and exports, and so forth. Both internal and exterior aspects of these elements exist. While businesses maintain certain policies for themselves, certain laws impact the business climate in each nation. These perspectives are considered in legal analysis, which subsequently lays out the tactics in light of these laws. For instance, consumer protection legislation, safety regulations, labor laws, and so forth.
6. **Environmental factors:** These factors include all those that influence or are determined by the surrounding environment. Factors of a business environmental analysis include but are not limited to climate, weather, geographical location, global changes in climate, environmental offsets, ground conditions, groundwater contamination, nearby water sources, and so forth.

3.5.1 Benefits of the PESTLE Analysis

1. Understanding the corporate environment through PESTLE analysis helps to discover opportunities and address threats.
2. It helps the organization comprehend the environment in which it operates.
3. It enables the business to use data strategically and create goals to support realizing the organization's mission and vision.
4. Analysis enables the organization to pinpoint important influences and forces, create action plans, and establish performance improvement objectives.
5. It produces data for the organization to help them evaluate and plan its operations for better performance.
6. The instrument makes it possible to evaluate the current environment and potential changes.
7. It aids in information gathering, analysis, and utilization to enhance organizational performance.

3.6 RISK MONITORING AND CONTROL

Implementing risk response plans, monitoring identified risks, tracking residual risks, detecting new risks, and gauging the efficiency of risk management procedures is known as controlling risks. The principal advantage of this method is that it enhances the effectiveness of the risk approach across the project life cycle to constantly optimize risk responses. Throughout the course of a technological project, planned risk responses that are included in the risk register are put into action, but the work being done on the project should be regularly inspected for new, evolving, and outdated risks. The control risks process uses methodologies that call for the use of performance data produced during project execution, such as variance and trend analysis. The control risks processes also seek to ascertain if

1. the project's presumptions remain true,
2. analysis reveals that a risk that was assessed has altered or can be retired,
3. policies and procedures for risk management are being followed,
4. cost or schedule contingency reserves should be adjusted in accordance with the most recent risk analysis.

Choosing other tactics, carrying out a backup plan, taking remedial action, and revising the project management plan are all examples of ways to control risks. The risk response owner regularly updates the project manager on the plan's success, any unexpected impacts, and any necessary adjustments to address the risk appropriately. Control risks also include updating the organizational process assets, such as project lesson-learned databases and risk management templates to benefit upcoming projects. Change requests can arise when emergency plans or workarounds are implemented. Change requests may also contain suggested remedial and preventative measures. In accordance with the most recent risk assessment, contingency reserves for cost or schedule should be changed. The related component documents of the project management plan are changed and reissued to reflect the approved modifications if the accepted change requests have an impact on the risk management procedures.

3.7 SUMMARY

High degrees of risk and complexity are present in the global corporate environment, which is a prerequisite for further growth and development. In particular, managers must handle various risks linked to technology, finance, insurance, environmental safety, and regulations. As a result, risk management is essential in many business domains that affect profitability, efficiency, and sustainability.

Due to the intricacy of risk management, businesses must have a written policy for project risk management and supplementary analytical tools. These analytics comprise risk identification tools such as brainstorming, checklists, influence diagrams, cause-and-effect diagrams, risk analysis tools such as probability and impact grids, event tree analysis, sensitivity analysis and simulation, Delphi techniques, expert judgment, risk response tools such as influence-predictability matrix, risk response planning chart, project risk response planning, and effective risk monitoring and control.

REFERENCES

Ahmadi-Javid, A., Fateminia, S.H., and Gemünden, H.G. (2020). A Method for Risk Response Planning in Project Portfolio Management. *Project Management Journal* 51(1), 77–95. https://doi.org/10.1177/8756972819866577.

Al Ariss, A., and Guo, G.C. (2016). Job Allocations as Cultural Sorting in a Culturally Diverse Organizational Context. *International Business Review* 25(2), 579–588.

Alsaadi, N., and Norhayatizakuan, N. (2021). The Impact of Risk Management Practices on the Performance of Construction Projects. *Esudios de Economia Aplicada* 39(4). http://dx.doi.org/10.25115/eea.v39i4.4164.

Aven, T. (2016). Risk Assessment and Risk Management: Review of Recent Advances on Their Foundation. *European Journal of Operational Research* 253(1), 1–13. https://doi.org/10.1016/j.ejor.2015.12.023. ISSN 0377–2217.

Bahamid, R.A., and Doh, S.I. (2017) A Review of Risk Management Process in Construction Projects of Developing Countries. *IOP Conference Series: Materials Science and Engineering*, 271, Article ID: 012042. https://doi.org/10.1088/1757-899X/271/1/012042.

Buye, R. (2021). Critical Examination of the PESTEL Analysis Model. Project: Action Research for Development. *Research Gate*. https://www.researchgate.net/publication/349506325_Critical_examination_of_the_PESTEL_Analysis_Model.

Christopher, P. (2017). The Risk Matrix: Uncertain Results? *Policy and Practice in Health and Safety* 15(2), 131–144. https://doi.org/10.1080/14773996.2017.1348571.

Coccia, M. (2018). The Fishbone Diagram to Identify, Systematize and Analyze the Sources of General Purpose Technologies (January 11, 2018). *Journal of Social and Administrative Sciences* 4(4), 291–303. https://ssrn.com/abstract=3100011.

Dario, P. (2017). *Risk Management in Construction Projects: A Knowledge Management Perspective From Swedish Contractors*. Master Thesis. Royal Institute of Technology, Sweden.

David, L.O., Adepoju, O., Nwulu, N., and Aigbavboa, C. (2023). Determining the Impact of Economic Indicators on Water, Energy, and Food Nexus for Sustainable Resource Security. *Clean Technologies and Environmental Policy* 26, 803–820. https://doi.org/10.1007/s10098-023-02651-8.

Evrin, V. (2021). Risk Assessment and Analysis Methods: Qualitative and Quantitative. *ISACA Journal* 2.

Gajewska, E., and Ropel, M. (2011). *Risk Management Practices in Construction Project: A Case Study*. Master of Science Thesis. Chalmers of Technology, Goteborg.

Garnett, J. (2018). *Quantitative Risk Analysis and Monte Carlo Simulation*. Paisely (Renfrewshire): The University of the West of Scotland.

Garrido, M.C., Ruotolo, M.C.A., Ribeiro, F.M.L., and Naked, H.A. (2011). Risk Identification Techniques Knowledge and Application in the Brazilian Construction. *Journal of Civil Engineering and Construction Technology* 2(11), 242–252.

George, C. (2020). The Essence of Risk Identification in Project Risk Management: An Overview. *International Journal of Science and Research (IJSR)* 9(2), 973–978. https://www.ijsr.net/archive/v9i2/SR20215023033.pdf.

Greenberg, M. (2017). *Explaining Risk Analysis: Protecting Health and the Environment*. New York, NY: Earthscan/Routledge.

IOSH (2018). *IOSH Managing Safely*. Wigston, United Kingdom: Institution of Occupational Safety and Health.

International Electrotechnical Commission/International Organization for Standardization (IEC/ISO). (2019). *IEC 31010 Risk Management—Risk Assessment Techniques*. Geneva, Switzerland: International Standard Organization. https://www.iso.org/obp/ui/en/#iso:std:iec:31010:ed-2:v1:en,fr.

Jayasudha, K., and Vidivelli, B. (2014). A Study on Risk Assessment in Construction Projects. *International Journal of Modern Engineering Research (IJMER)* 4(9), 20–23.

Liu, J., Zhao, X., and Yan, P. (2016). Risk Paths in International Construction Projects: Case Study From Chinese Contractors. *Journal of Construction Engineering and Management* 142, 05016002.

Owojori, A.A., Akintoye, I.R., and Adidu, F.A. (2011). The Challenge of Risk Management in Nigerian Banks in the Post Consolidation Era. *Journal of Accounting and Taxation* 3(2), 23–31. https://academicjournals.org/journal/JAT/article-full-text-pdf/E163392746.

Peace, C. (2017). The Risk Matrix: Uncertain Results? *Policy and Practice in Health and Safety* 15, 1–14. https://doi.org/10.1080/14773996.2017.1348571.

PMI (2008) *A Guide to the Project Management Book of Knowledge (PMBOK)*. 4th Edition. Newtown Square: Project Management Institute.

Project Management Institute. (2013). *A Guide to the Project Management Body of Knowledge (PMBOK® Guide)*. 5th Edition. Newtown Square: Project Management Institute.

Rastogi, N., and Trivedi, M.K. (2016). PESTLE Technique—A Tool to Identify External Risks in Construction Projects. *International Research Journal of Engineering and Technology (IRJET)* 3(1), 384–388.

Rehacek, P. (2017). Risk Management Standards for Project Management. *International Journal of Advanced and Applied Sciences* 4(6), 1–13.

Rehacek, P., and Bazsova, B. (2018). Risk Management Methods in Projects. *Journal of Eastern Europe Research in Business and Economics*, 1–11.

Renault, B.Y., Agumba, J.N., and Ansary, N. (2016). A Theoretical Review of Risk Identification: Perspective of Construction Industry. In: Mojekwu, J.N., Nani, G., Atepor, L., Oppong, R.A., Adetunji, M.O., Ogunsumi, L., Tetteh, U.S., Awere, E., Ocran, S.P., and Bamfo-Agyei, E. (eds.) *Proceedings of 5th Applied Research Conference in Africa (ARCA) Conference*, August 25–27, 2016, Cape Coast, Ghana.

Rostami, A. (2016). Tools and Techniques in Risk Identification: A Research Within SMEs in the UK Construction Industry. *Universal Journal of Management* 4(4), 203–210. https://doi.org/10.13189/ujm.2016.040406.

Said Djamaluddin, S., and Herawati, A. (2020). The Application of Risk Management Practices at PT XYZ: Case Study at Industrial Estate Company in 2018. *International Journal of Innovative Research & Development*. https://doi.org/10.24940/ijird/2020/v9/i1/JAN20020.

Sammut-Bonnici, T., and Galea, D. (2015). PEST Analysis. https://doi.org/10.1002/9781118785317.weom120113.

Sotic, A., and Rajic, R. (2015). The Review of the Definition of Risk. *Online Journal of Applied Knowledge Management* 3(3), 17–26. https://www.iiakm.org/ojakm/articles/2015/volume3_3/OJAKM_Volume3_3pp17-26.pdf.

Srinivas, K. (2019). Process of Risk Management. In: *Perspectives on Risk, Assessment and Management Paradigms*. London, UK: IntechOpen. https://doi.org/10.5772/intechopen.80804.

Tworek, P. (2012). Plan Risk Response as a Stage of Risk Management in Investment Projects in Polish and U.S. Construction—Methods, Research. *Analele Ştiinţifice ale Universităţii» Alexandru Ioan Cuza «din Iaşi. Ştiinţe economice* 59(1), 201–212. https://doi.org/10.2478/v10316-012-0014-9.

Zidafamor, E. (2016). Risk Management in Nigerian Financial Institutions—A Literature Review. *Research Gate*. https://www.researchgate.net/publication/301891410_Risk_Management_in_Nigerian_Financial_Institutions_-_A_Literature_Review.

Part II

Risks of the Fourth Industrial Revolution Technologies

4 Artificial Intelligence

4.1 INTRODUCTION TO AND COMPONENTS OF ARTIFICIAL INTELLIGENCE

The study of intelligent machines that function and behave like people is the focus of the computer science subfield known as artificial intelligence (AI). The "I" in "AI" stands for machine intelligence or the way that computers apply algorithms. The development of all human technologies and evolution is a result of intelligence. Mumbai (2014) asserts that it was held true that human attempts to learn to represent a violation of the laws of God or nature. Even in Greek mythology, for instance, in Prometheus, artificial intelligence has a long history that begins with Aristotle. He carefully examined the visions, wonders, and uncertainties about nature and codified them into a structured philosophy. He believed that the fundamental source of all knowledge was the study of thought. At the Dartmouth Conference in 1956, researchers from the Massachusetts Institute of Technology (MIT), IBM, and Carnegie Mellon University (CMU) came together to launch the field of artificial intelligence research. AI is achieved by researching how the human brain works, learns, reacts, and works while attempting to solve a problem and by using the findings of this research as the foundation for creating intelligent software and systems. After the invention of the computer and machines, humans developed the power of computer systems due to their various uses, growing speed, and decreasing size over time (Tutorials Point 2015).

"The science and engineering of constructing clever machines, especially intelligent computer programs," is how John McCarthy, the pioneer of artificial intelligence, described AI (McCarthy 2007). He suggests that while there are many issues that humans and other living things can address, AI does not yet have a solid algorithm to support it. Tegmark (2017) defined artificial intelligence as a (software and hardware) system designed by humans and assumed as a complex goal, acting in the physical or digital realm by perceiving their environment through data acquisition, interpreting the collected structured or unstructured data, reasoning on the knowledge, or processing the information, derived from the data, and deciding the most effective action to achieve the goal. Artificial intelligence is a method for teaching a computer-controlled robot or piece of software to reason intelligibly, much like a typical human might.

Howie (2015) claimed that despite how robots with human-like qualities are portrayed in science fiction, AI includes everything from Google's search algorithms to IBM's Watson to autonomous weapons. The term "narrow AI" (sometimes known as "weak AI") refers to the type of artificial intelligence used today (facial recognition, Internet searches, or only driving a car). However, many researchers are working toward developing general AI, AGI (artificial general intelligence), or ANI in the long run (artificial natural intelligence).

DOI: 10.1201/9781003522102-6

Artificial intelligence is the capacity for calculation, reasoning, perception, information storage and retrieval from memory, learning, linguistic proficiency, problem-solving, and adaptation (Mumbai 2014). Here is a list of the features of AI, that are demonstrated by a computer or a system or an application:

- Reasoning
- Learning
- Problem-solving
- Perception
- Linguistic intelligence

Reasoning: The set of processes enables us to provide the basis for conclusions, decisions, and prediction. It is usually viewed from two perspectives of inductive reasoning and deductive reasoning. Inductive reasoning conducts specific observations in broad general statements, whereas deductive reasoning entails making general statements and examining the possibility of reaching a specific and logical conclusion.

Learning: It is the process of learning something new through study, practice, instruction, or personal experience. Learning increases knowledge of the study topics. Humans, animals, and AI-enabled systems all have an impact on learning, which is classified as the following:

- **Auditory learning:** It is learning by listening and hearing, for example, a musician listening to his recorded music albums.
- **Episodic learning:** To learn by remembering sequences of events one has witnessed or experienced. This is linear and orderly.
- **Motor learning:** It is learning by the exact movement of muscles, for example, picking objects, writing, weightlifting, and so forth.
- **Observational learning:** To learn by watching and imitating others, for example, a child trying to learn by mimicking her parent.
- **Perceptual learning:** It is learning to recognize stimuli that one has seen before, for example, identifying and classifying objects and situations.
- **Relational learning:** It involves learning to differentiate among various stimuli based on relational properties rather than absolute properties.
- **Spatial learning:** It is learning through visual stimuli such as images, colors, maps, and so forth, for example, a person creating a road map in mind before actually following the road.
- **Stimulus-response learning:** It is learning to perform a particular behavior when a certain stimulus is present. For example, a dog raises its ear on hearing a doorbell.

Problem-solving: It is the process through which one observes and attempts to overcome a current predicament by following a path obstructed by known and unknowable barriers. Making decisions as part of problem-solving entails choosing the most appropriate option from a variety of accessible options to achieve the desired result.

Perception: It is the procedure for gathering, analyzing, prioritizing, and arranging sensory data. Sensory organs contribute to perception. In the field of artificial intelligence, perception mechanisms combine sensor data in a meaningful way.

Linguistic Intelligence: It pertains to one's capacity for verbal and written language use, comprehension, expression, and composition. Interpersonal communication requires it (Mumbai 2014).

4.2 RISK ASSOCIATED WITH ARTIFICIAL INTELLIGENCE

To surpass human intellect in every aspect, artificial intelligence has been widely developed into AGI and ANI in the 21st century (Barrett and Baum 2017). Google's Deep Mind, Facebook's facial recognition technology, Apple's Siri, Amazon's Alexa, Tesla, and Uber's self-driving cars are just a few examples of existing ANI systems (Stanton et al. 2020). An AGI would have a higher intelligence (Bostrom 2014) and the capacity to complete complicated tasks in challenging circumstances (Goertzel 2006). AI is predicted to pose significant risks to the global economy, physical health, and social welfare (Vallor 2017; David et al. 2023). AI is a potent catalyst that is helpful, harmful, or disruptive to socioeconomic forces. Many researchers have analyzed that there is likely to be widespread explosion that could affect the upcoming era and generations to come. This explosion could be caused by the exponential rate at which technology evolves, such as in computing power, data science, neuroscience, and bioengineering (Kurzweil 2005; Dogo et al. 2019; Olukanmi et al. 2022; David et al. 2022; Adepoju et al. 2022). In addition to displacing the workforce and manipulating political and military systems, the development of AGI, derived from AI, could result in extinction, ending the human species (Bostrom 2002; Sotala and Yampolskiy 2015). Elon Musk, the founder of Tesla and SpaceX, claims that artificial intelligence is "much more destructive than nukes and could relatively kill small portions of the earth," similar to pandemics "and thought to be virulent." The risks are as follows:

1. **Technology troubles:** The performance of AI systems can be significantly impacted by technological issues that affect the overall operational environment of a technological project, because the data feeds are no longer included in the customer trades. For instance, a financial institution faced significant difficulties after its compliance software failed to identify trading concerns.
2. **Communication issues:** A key area where there is risk is when a machine (AI) interacts with a human, whereby the interpretation of the result from what is fed to the system is flawed. Hence, risks occur when communication lapses or outcome from wrong input influences decision-making. The automated transportation, industrial, and infrastructure systems provide notable interface concerns. These applications, for instance, self-driving automobiles, run the risk of causing accidents and injuries if operators of heavy machinery, vehicles, or other machinery

cannot recognize when systems are overridden or compromised because the operator's attention is diverted. On the other hand, human judgments can be flawed when interpreting what the AI technology says. In addition, inaccuracies in scripting, data management gaps, and model-training data errors can quickly jeopardize compliance, fairness, privacy, and security in data analytics. Also, for example, a sales team that is more adept at selling to a particular demographic may unwittingly educate an AI-driven sales tool to reject a particular group of consumers. Without strict controls, avaricious staff or hostile outside parties would be able to tamper with algorithms or utilize an AI program dishonestly.

3. **Black box algorithms and lack of transparency:** Many AI systems' main objective is forecasting; therefore, the procedures they use might be so intricately developed that even their designers are unable to explain how the different variables combined to get the final prediction exactly. This has made it a difficult task for monitoring and control, especially by regulatory agencies, in the interest of public good. These transparency issues constitute a risk, as it might be used for fraud, corporate espionage, aiding and abetting crime, and corrupt practices, among others. This has led legislative authorities to start considering what checks and balances need to be set up, which is why some algorithms are referred to as "black boxes." For instance, some businesses incur the risk of being unable to justify decisions, such as denying a banking account based on an AI prediction about the applicant's creditworthiness, due to how the AI system operates, thereby making some losses in the process (Mckeown 2021).
4. **Models misbehaving:** Artificial intelligence models, either via machine learning, deep learning or from artificial neural networks, have the tendency of misbehaving and functioning against expectation or project goals. For instance, a data item used in training an AI system may lack some information that makes the data incomplete, thus producing biased results, and making conclusions unbalanced. An example is someone being denied a job application because certain criteria fed into the system are against such a type of application, which in fact are needed. Therefore, this type of risk is the result of poor representation of data, inconsistent metrics, poor knowledge of situations and circumstances, and the lack of emotional intelligence in feeding the AI system. Cheatham et al. (2019) opined that when AI models are concealed in software-as-a-service (SaaS) products, they can be more challenging to detect, which gives them the high tendency to misbehave. For instance, when suppliers add new, intelligent features—often without much fanfare—they are also introducing models that could interact with the data of users in producing unexpected dangers, which hackers could exploit.
5. **Data sourcing and violation of personal privacy:** Businesses have access to enormous volumes of structured and unstructured data, and the International Data Corporation predicts that the world's data sphere will grow from 33 zettabytes (33 trillion gigabytes) in 2018 to 175 zettabytes

(175 trillion gigabytes) by 2025. This is an exponential increase in data, which will increase the propensity to increase customer data, for business decision, AI modeling, thus violating their personal privacy, as most times, the data source is with different objectives in variance with their utilization. Moreover, data leakages or breaches can adversely harm a company's brand and increase the possibility of legal infractions in an era where many legislative bodies have passed laws restricting the use of personal data. A well-known regulatory example is the General Data Protection Regulation (GDPR) adopted by the European Union in April 2016, which subsequently influenced the California Consumer Privacy Act passed in June 2018 (Mckeown 2021).

6. **Strategy risk:** This kind of risk entails deploying AI to alter competitors' business strategies, leading to poor decision-making and disorganized policies. This kind of risk also occurs between hostile nations when AI technologies are deployed against a nation, without the recipient nation having adequate knowledge about the deployment but seeing only the catastrophic outcome. A concise initiative on the items associated with AI should be known, together with how to improve good influence on people and solutions to things not going well, to develop AI strategy rather than relying on common knowledge (Caner and Bhatti 2020).
7. **Financial risk:** Due to the unexpected nature of AI work, AI/ML oriented projects should not be approached like other IT projects. This is because AI is more complicated and involves infrastructure expenditures, management and governance data, and human supervision. This makes the financial incursion a huge expenditure, thus making it riskier if AI oriented project does not achieve intended project goals. The financial situation of AI's fundamental component should be considered to avoid financial waste in terms of expenditures and insurance, both during and after execution. Bhatti (2020) opined that during AI project execution, the finance manager should be aware of this to avoid incurring high costs for the AI experiment, as it differs from other technological projects.
8. **Technical risk:** This is the most obvious risk, as organizations always tend to push through an immature and untested AI technology to the market to quickly gain market share or market dominance, which often results in technical issues. There are various technical risks: Information about data, including its appropriateness, representativeness, and inflexible data infrastructure. These technical risks also include model hazards, which entail weaknesses in performance or dependability in the real world. Also, this technical risk is reflected in concept drift, which is referred to changes over time brought on by unanticipated circumstances in the market or environment. Technical risks also include lack of DataOps, MLOps, IT team unprepared for ML/AI deployment or scaling requirements, security breaches, and IP theft. Other examples include implementation delays, errors, incomplete testing, a lack of stable or mature environments and a lack of stable or mature environments.

Working on team skills, investing in contemporary data infrastructure, and adhering to best practices for ML are the greatest ways to mitigate problems (Bhatti 2020).

9. **People and process risk:** Most times, the success of a project is dependent on the correct organizational structure, culture, and people. Hence, the success of an AI project depends on the developers' competency and a positive work environment, whereby failure to have them constitutes a risk to an AI project, which jeopardizes the AI process and procedures for optimal functionality. These risks make people conceal issues and shirk responsibility because the working climate is neither friendly nor open to discussion, not alone when incompetent people are in place. These risks will lead to the inability to grow AI due to a lack of operational understanding, restrictive attitude, miscommunications, outdated IT, process flaws, coordination problems, model dependence, insufficient human supervision, and feedback loops for learning weak technical grounding (Bhatti 2020).
10. **Trust and explainability risk:** People are usually reluctant to utilize or adopt a model even when you put a lot of effort into developing an AI-powered application, due to the issue of trust and lack of understanding of its performance. These risks are due to poor model performance under specific circumstances, opaqueness (lack of results explanation), lack of assistance when issues occur, poor user experience, lack of incentive alignment, and disturbance to people's workflow. Therefore, these risks lead to poor business performance, inaccuracy in decision-making, poor feedback mechanism from the beneficiaries, poor return on investment.
11. **Ethical risk:** Most times AI/ML projects have excellent data, an outstanding technical staff, obvious benefits, and no legal encumbrance or pitfalls, yet there is a question to be answered—is it ethical? This risk is narrowed down to issues of morality in deploying AI to achieve certain objectives, especially in terms of racial issues, unfair predictions, and biases in the allocation of resources. Also, ethical risks entail issues of the moral clause when AI jobs take over millions of people's jobs, or how can machine-created job ensure equitable wealth creation.
12. **Algorithmic risk:** These risks constitute the inappropriateness of data that is input in AI systems, biased logic and flawed assumptions in the design of the algorithms, coding errors, and spurious patterns in training data. Cases of algorithm risks were cited during the 2016 Brexit referendum, where algorithms were blamed for the 6% flash crash of the British pound for two minutes. According to Deloitte (2017), underlying factors causing algorithm risks include human biases, technical flaws, usage flaws, and security flaws. That is, the design of algorithms is susceptible to dangers like biased reasoning, improper modeling methodologies, incorrect presumptions or judgments, coding mistakes, and finding fictitious patterns in training data.

4.3 RISK MANAGEMENT TECHNIQUES FOR ARTIFICIAL INTELLIGENCE

Risk management is "a systematic way of looking at areas of risk and consciously determining how each should be treated. It is a management tool that aims at identifying sources of risk and uncertainty, determining their impact, and developing appropriate management response" (Srinivas, 2019). Risk management is a vital activity implemented in the organization for the protection of value and improvement of performance in a wide range of areas, including health and safety, compliance, environmental protection, the efficiency of operations, governance, and reputation to achieve its stated objectives (A.S.R.M 2009). Risk management is the process that allows IT managers to balance the operational and economic costs of protective measures and achieve gains in mission capability by protecting the IT systems and data that support their organizations' missions (NIST 2002). It is also described as the coordinated activities used in directing and controlling the organization regarding the outcome of uncertainty or risk. Machine learning techniques are a core establishment of artificial intelligence, which was transformed and developed to approach financial risk management. In the late 1990s machine learning (ML) was very popular and is effective in building intelligent systems capable of making decisions without explicit programming (Dua and Du 2011; Dogo et al. 2019).

Ngai et al. (2011) provide an excellent overview of the core AI techniques used in financial fraud detection and note the main techniques applied, such as decision trees and neural networks. Further emphasizing, AI's ability to introduce better process automation can accelerate the pace of routine tasks, minimize human error, process unstructured data to screen out relevant content or negative news, and determine individuals' connectedness to evaluate risky clients and networks. The Federation of European Risk Management Associations (2019) described machine learning as the methodology for implementing algorithms and statistical models that computer systems use to perform a specific human task effectively without explicit instructions to the machine.

Sanford and Moosa (2015) opined that in the risk management process, artificial intelligence could assist organizations at various stages, ranging from identifying risk exposure to measuring, estimating, and assessing its effects. It can also help opt for an appropriate risk mitigation strategy and find instruments to facilitate shifting or trading risk. Everything associated with regulating risk is driven through the growth of AI-driven solutions: from deciding how much a bank should lend to a customer, to providing warning signals to financial market traders about position risk, to detecting customer and insider fraud, and improving compliance and reducing model risk (FSB 2017).

However, just as artificial intelligence is used in managing risks, there are several categories of risk management that can be deployed to manage the risks associated with artificial intelligence. These are as follows.

4.3.1 General Risk Management Process

4.3.1.1 Risk Identification

Risk identification for an artificial intelligence–oriented project is a two-part process involving detecting risks that can be inappropriately allocated and documenting their possible time and cost impacts. In this technique, the organization needs to identify the actual and potential risk associated with assets valuation and services rendered; hence, the organization must develop and calculate a comprehensive list of tangible and non-tangible assets based on their level of importance. Factors to be considered include an asset's financial value and the data stored in the asset. It is essential that people with appropriate knowledge of the business are involved in identifying risks. The AI teams who are delegated for the work must be capable and expert in the field to identify the areas where they are exposed to risk, the extent and severity of the risk, and the frequency of the occurrence which gives rise to the risk (Australian Standard on Risk Management 2009). FERMA (2019) connotes that each IT asset needs to be affiliated with business procedures they support through underlying databases, applications, operation systems, networks, hardware, people, locations, and so forth. Manual processes should also be considered since hardcopy documents (such as contracts) can contain confidential information. Hardcopy documents are also information and tangible assets, as they form the backbone of AI projects regarding objectives, functions, and operations.

Tom et al. (2018) opined that the organizations must identify the risks which could have a material adverse impact on the organization's business strategy and operations. It was also stated that organizing meetings across the organization with risk management, AI users, data scientists, and its teams could comprehend AI risks. The purpose of identifying risks is to obtain a list of risks that have the potential to have a cascading effect on the progress of the project, and different techniques are applied to manage/mitigate them. To find all potential risks which might impact a specific project, different techniques are applied (Srinivas 2019).

4.3.1.2 Risk Assessment

An organization should design and agree upon a risk assessment process before developing each AI use case. Embedding a risk assessment process could bring about assessing the level of risk exposure in the businesses or project reputation and implications (Tom et al. 2018). The organization also needs to understand the required output of the risk assessment, that is, what is likely to impact or affect the terms required for strategy and business objectives. The organization's techniques for accessing risk are either qualitative or quantitative (ISO 2009). Qualitative risk assessment is subjective to the assessor's experience and other factors. Meanwhile, quantitative risk assessment is objective and based on facts and figures. Identifying and assessing IT assets is essential because of their vulnerability against cybercrime (hackers and fraudsters).

4.3.1.3 Risk Controlling

In every business organization or AI-oriented project environment, controls are risk mitigating, countermeasures, processes, and rules that need to be instigated to obtain a clearer view of actual risk levels and to function in alignment with management intention and external regulations. An organization's control process tends to deliberate how the risks of AI will be felt by stakeholders (customers, underwriters) and what the touch points are, as controls are not continually operating effectively; therefore, audit procedures should be in place to assess the effectiveness and for more developed companies, control efficiency. Relevant controls are likely to span multiple areas (e.g., HR, technology, operations) due to the widespread adoption of AI.

McLean et al. (2021) posit that in curbing risk for an AI, specific controls need to be induced, which include programming, development, increased data-driven modeling, improved risk management processes, international regulations, government control, taxation of the AGI, and consultation with a wide range of experts, among others. Salmon et al. (2021) opined that the organization must ensure risk control by creating safe artificial general intelligence and effectively managing associated risks as they progress. ISO/IEC (2011) stated that inherent risks could be alleviated with a control basis since inherent risks are risks calculated from the business criticality of an IT asset, the likelihood of relevant threat occurrence, and the vulnerability of the IT asset. Security controls comprise technical and nontechnical methods for risk contingency for information technology. Technical controls are safeguards incorporated into computer hardware, software, or firmware (e.g., access control mechanisms, identification, authentication mechanisms, encryption methods, and intrusion detection software). Nontechnical controls are management and operational controls, such as security policies, operational procedures, and personnel, physical, and environmental security.

4.3.1.4 Risk Evaluation

Risk evaluation is a technique conducted to decide whether the risk is acceptable or unacceptable with the specific purpose of making decisions about future actions. Evaluating risk in response to an organizational level is to understand the overall impacts of the risk profile. Once risks are calculated, a priority list must be established to ensure the highest risks are controlled. It is also essential to compare the risks against the risk acceptance criteria of the organization. The risk evaluation level is estimated to reveal the potential and actual consequences that might arise from identified risks and should cease any activity that creates unnecessary risks or uncertainty. Once risks have been assessed, the severity and occurrence of risk should be examined and calculated. The tools used in assessing risk are probability, scenario planning, simulations, including Monte Carlo spreadsheet simulation, decision trees, real options modeling, sensitivity analysis, risk mapping, statistical inference, SWOT or PESTLE analysis, root cause analysis, cost-benefit/risk-benefit analysis, and human reliability analysis. Risk evaluation is estimated using quantitative and qualitative risk assessment and is further explained in the following.

Qualitative risk assessment: SBS CyberSecurity (2018) opined that qualitative risk assessment is studying an event or regulatory control and understanding the quality of its implementation. Qualitative risk assessment gives the risk evaluator and the risk manager information about how well the control has been instigated, usually done through non-numeric analysis, but on qualitative measures such as expert opinion, focus group discussion, interview survey among employees, and Delphi technique. Organizations employ qualitative assessment techniques to identify risk because an expert opinion is the best source, rather than an unreliable quantity.

Quantitative risk assessment: A quantitative risk assessment focuses on quantifiable and pre-defined data; the detailed quantitative evaluation of risk is the one that is identified as risk analysis. Quantitative assessment quantifies the potential impact of risks in terms of time, cost, and quality. The quantitative method in determining risk constraints requires analyzing historical data through statistical analysis.

4.3.1.5 Monitoring and Reporting of Risk

After identifying, assessing, evaluating, and controlling a risk, an appropriate monitoring and reporting system needs to be established to ensure that the treatment has been effective and the risk has been controlled (Raaghieb et al. 2017). Monitoring risk is a continuous activity that increases awareness of what is happening across different parts of the organization. Over time, monitoring risk enables management to identify critical trends, business opportunities, or process improvements. Reporting risk is connecting real-time risk and performance data to different stakeholders. An organization should be able to state its clear and accurate success metrics and key performance indicators (KPIs) and look for key risk indicators (KRIs) that might generate a change in strategy and monitoring of algorithms. In monitoring risk, the measure used must be included in the organization's fairness and anti-discrimination policy. Furthermore, operational monitoring should consist of capturing and comparing metrics such as the volume of transactions being pushed to a human underwriter by the AI system and the speed with which policies are written when the AI solution is organized relative to non-AI systems (Tom et al. 2018). Risk monitoring is necessary to ensure the execution of the risk plans and to evaluate their effectiveness in reducing risk.

4.3.2 Algorithmic Risk Management Framework

Continuous monitoring of AI algorithms, potentially with the assistance of other algorithms, is essential for effective control of algorithmic risks. To efficiently handle algorithmic hazards, organizations must update outdated risk management frameworks. Deloitte (2017) suggests establishing and implementing innovative strategies aligned with best practices and legal standards, grounded in a robust enterprise risk management foundation. This involves developing governance approaches in areas such as accountability and responsibilities, regulatory compliance, inventory, and risk classifications, disclosure to users and stakeholders, standardized principles, policies, and guidelines, as well as engagement in lifecycle and change management and the establishment of inquiry and complaint procedures, along with hiring and training of personnel. Additionally, as outlined by Deloitte (2017), risk management in the design, development, deployment, and use of algorithms should adhere to a standardized algorithm design process, employ data assessment procedures, define sets of assumptions and limitations, embed security and operations controls in the algorithms and deployment process. Furthermore, the algorithm requires monitoring and consistent testing, encompassing output logging and analysis, sensitivity analysis, continuous improvement, and independent validation.

Furthermore, this approach elements are discussed in the following.

Strategy and governance: This entails setting up a governance approach for an organization before the deployment of an AI technology, which should include values, guidelines, expectations, positions, and duties, among other things, which will control the setup of an algorithm. More specifically, it entails offering openness and procedures that let a company employ algorithms effectively, ethically, and in conformance with the law of the land.

Design, development, deployment, and use: An algorithm's choice and decision-making process should be based on defined processes, quality data source, tested assumptions and limitations, and the security guidelines for managing the algorithm. This should be seen from the algorithm life cycle, data selection to algorithm design, integration to actual live use in production, development processes, and procedures that align with an organization's governance structure.

Monitoring and testing: The organization must create procedures for evaluating and monitoring algorithm data inputs, operations, and outputs through necessary algorithm testing, conducting periodic sensitivity analysis, utilizing independent validation, and engaging in continuous process improvement. This stage entails that the organization must not be complacent with the performance of an algorithm but should engage continuously for quick identification of issues and risks and prompt deployment of necessary management mechanisms.

4.3.3 The Basel Framework

This framework is mostly for mitigation and protection against financial fraud that are occasioned by high level AI systems across regions. The Bank of International Settlements (BIS) created the Basel framework as a collection of international banking regulation norms to support the stability of the financial markets.

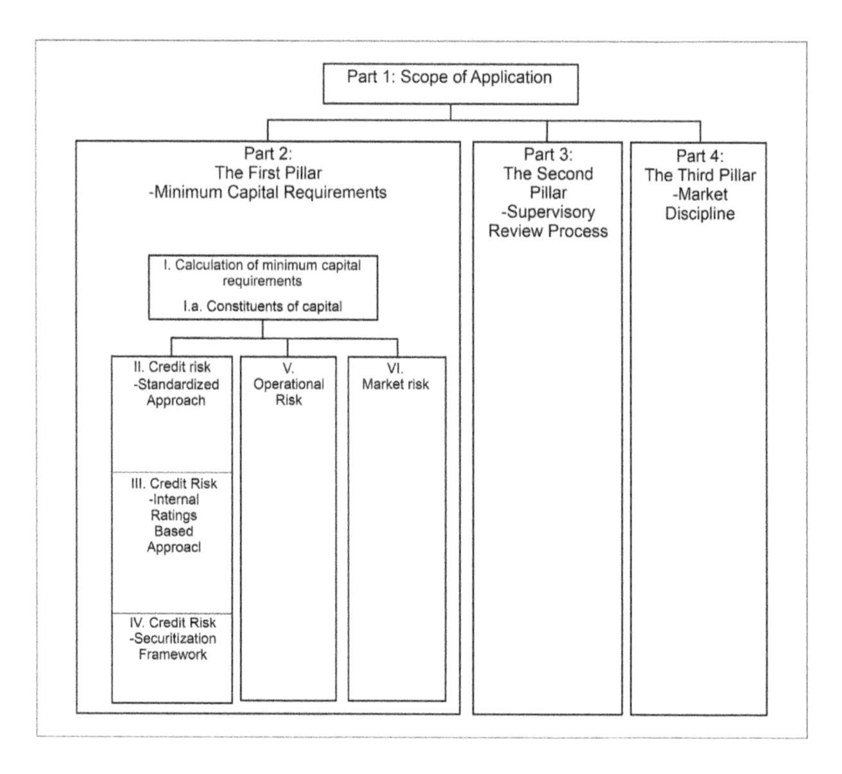

FIGURE 4.1 Basel II structure (So 2020, para 7)

BIS lacks regulatory authority, but because it serves as the "central bank of central banks," Basel norms are considered the norm worldwide. The Basel Committee on Banking Supervision (BCBS), which drafted the guidelines, was established during a global financial crisis. It was founded in 1974 by a group of 10 governors of central banks, and it today has 45 members from 28 different countries. Basel, for instance, sets standards for the development and use of algorithms in banks, where most AI powered financial fraud is concluded. This framework should be adopted to protect AI-oriented projects' financial assets. The Basel II framework can be seen in Figure 4.1 (So 2020).

4.3.4 Commercial Risk Management Framework

Commercial Risk Management Framework (CRMF) is a special-purpose business ecosystem that extends along supply chains and over single organization boundaries. This AI-oriented risk management framework is mostly recommended for small and medium-scale enterprises who use AI systems and solutions to solve business issues and execute projects. It tries to identify, manage, and reduce commercial risk through the collaborative efforts of parties with similar corporate objectives. As shown in Figure 4.2, this structure is anticipated to be used outside of any specific company, while all interested participant companies are given access to solutions for services (Zigiene et al. 2019).

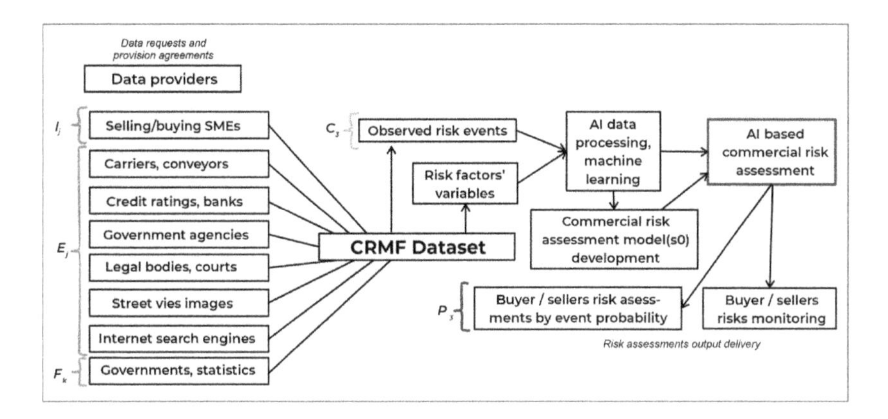

FIGURE 4.2 The ecosystem approach to commercial risk management framework (Zigiene et al. 2019, p. 12)

4.4 RISK MANAGEMENT SKILLS FOR ARTIFICIAL INTELLIGENCE

Analytical skills: Analytical skills are essential skills a risk manager must possess for analyzing and collecting data to assess and predict risk relating to artificial intelligence. Extensive research analysis of data is conducted to find the potential gap an artificial intelligence could use to improve. An eye for detail and sharp analytical skills are needed for AI to improve and develop. Analytical skills are imperative for AI for data gathering and database systems because, without data, there is no algorithm for AI to function effectively.

Problem-solving: If potential risks are found, one must be curious to look at the business issue and get to the crucial solutions. This entails the capability to think on one's feet and make adequate analysis and proffer solutions for the short, medium and long term, together with the potential impact of those solutions. Problem-solving is a skill one should possess in tackling AI risks, which is the activation of one's mental and cognitive ability to quickly mitigate or put a stop to the effect before snowballing.

Communication skills: Communication skills are essential for people in sales, customer service, information technology companies, human resources, and so forth. Solid interpersonal and communication skills are vital for risk managers to interface with AI risk. To keep updated on risk management best practices, risk managers always need to research, read, and network. Artificial intelligence (AI) can help train communication skills in new methods. Speech and language technology can be created for learners to practice communication skills. AI algorithms can automatically assess communication skills and provide feedback and personalized training. Communicating effectively includes knowing and using credible language to explain ideas clearly, and declaratively with evidence (Alelo 2019). Further emphasized by (Alelo 2019), it was asserted that a new technology called Enskill EPC (effective and persuasive communication), developed by Alelo, could help communication perform effectively in a way that learners who present structured arguments will be aided by presenting their vision, explaining their strategy, and then proposing guide to settle the argument. Using AI technology, Enskill EPC automatically detects elements of an argument and ensures they come together well-structured. Risks managers can also use this in a technological project, as this is essential in communicating with other members of the team or appropriate risk owner or authority on the occurrence of a certain AI risk. Communication can be verbal, which must be fluent and easy to understand; it can also be written, which must be simple to understand, as vague words or technical jargon confuses an individual during the implementation of a suitable risk framework.

Networking skills: Risk managers should constantly connect with other professionals in the fast-paced, quickly shifting financial world. Using technology and social networks such as LinkedIn and Twitter, together with the associated communities, professionals can develop connections with other professionals worldwide. Moreover, several financial centers, such as London, host several Risk Management Committees that organize periodic conferences and lectures about risk management and regulation, which can be a great place to meet other risk management professionals. There are numerous advantages of networking:

- Increase awareness and knowledge of risk management.
- Broaden the risk manager's mind-set, helping them understand and prioritize what topics are relevant.
- Help develop the risk management "lingo," keeping up with current topics and contributing to increasing their knowledge.
- Be exposed to initiatives taking place and participate in them. Being willing to go the extra mile will benefit you as a professional and help increase your knowledge.

Strategic thinking: Solid risk managers must be forward looking and strategic minded, able to understand potential risks for the firm at the departmental level and from a broader firm perspective. The head of risk management must keep pace with technology's transient and volatile nature, as there are new ways of circumventing technological security every day, thus requiring strategic thinking. While controlled by nature, their function should also assist the front officer staff and senior management in delivering solutions that align with business objectives and obey risk safeguards.

Furthermore, there are technical skills for risk management in artificial intelligence, which include and are not limited to the following:

Programming languages: Every AI enthusiast must have sound knowledge of some programming languages such as Artificial Intelligence Markup language, Prolog, Haskell, Julia Java, R, Python, C++, and JavaScript. These languages can be learnt online and several tutorial videos exist. In managing the risks of AI, knowledge about this programming language is essential, as they give you hindsight and foresight about potential risks.

Mathematical knowledge: AI is built on algorithms and applied mathematics. Therefore, strong analytical and problem-solving skills and sound mathematical knowledge are required to solve problems in artificial intelligence.

Machine learning: This is a subset of artificial intelligence. Its algorithms can train a machine to learn without further programming. Hence, knowledge of machine learning is needed in artificial intelligence as it helps AI systems behave like humans (Pathak 2022).

Deep learning: This is a branch of machine learning and data science. It develops systems that have the predictive capability and statistical inference.

Shell scripting: This is a core element of artificial intelligence that recognizes the pattern. As AI development advances, it will get to a point where data processing will be carried out on Linux-based machines. Hence, it will be challenging to work with AI functions when the shell command functions are not known (Pathak 2022).

Clustering analysis: This skill is used in software development, anomaly detection, crime analysis, risk analysis, education, information retrieval, and so forth. Clustering analysis is the act of grouping objects with clustering algorithms.

4.5 SUMMARY

This study focuses on the introduction and components of artificial intelligence, the risks associated with artificial intelligence, the risk management technique in artificial intelligence, and the risk management skills for artificial intelligence. The artificial intelligence components discussed in this chapter include heuristic problem-solving, learning, reasoning, perception, and linguistic intelligence. The associated risk may include technology troubles, communication issues, black-box algorithms and lack of transparency, models misbehaving, data sourcing and violation of personal privacy. Furthermore, risk management techniques were discussed to mitigate this associated risk in AI. The techniques discussed in this chapter include risk identification, risk assessment, risk controlling, risk evaluation, monitoring, and reporting of risk, which all form the general risk management approach. Inclusively, the risk management techniques in AI discussed in this study are algorithmic risk management framework, Basel framework, and CRMF. Furthermore, the soft skills needed for risk management in AI include analytical, problem-solving, communication, networking, strategic thinking, and so forth. The technical skills required for risk management in AI include the following: programming languages (such as python, Java, R, JavaScript), mathematical knowledge, machine learning, deep learning, shell scripting, and clustering analysis.

REFERENCES

Adepoju, O., Aigbavboa, C., Nwulu, N., and Olaiya, M. (2022). *Reskilling Human Resources for Construction 4.0. Implications for Industry, Academia, and Government.* Springer. https://doi.org/10.1007/978-3-030-85973-2.

Alelo. (2019). Communication Skills for Artificial Intelligence. www.alelo.com/2019/11/how-ai-simulations-close-the-communication-skills-gap/#:~:text=Artificial%20intelligence%20(AI)%20can%20help,provide%20feedback%20and%20personalized%20training.

Australian Standard on Risk Management. (2009). NSW Government, Family & Community Services. AS/NZS ISO 31000:2009.

Barrett, A.M., and Baum, S.D. (2017). A Model of Pathways to Artificial Superintelligence Catastrophe for Risk and Decision Analysis. *Journal of Experimental and Theoretical Artificial Intelligence* 29(2), 397–414. https://doi.org/10.1080/0952813X.2016.1186228.

Bhatti, B. (2020). 7 Types of AI Risk and How to Mitigate Their Impact. *Medium.* https://towardsdatascience.com/7-types-of-ai-risk-and-how-to-mitigate-their-impact-36c086bfd732.

Bostrom, N. (2002). Existential Risks: Analyzing Human Extinction Scenarios and Related Hazards. *Journal of Evolution Technology* 9. https://ora.ox.ac.uk/objects/uuid:827452c3-fcba-41b8-86b0-407293e6617c.

Bostrom, N. (2014). *Superintelligence: Paths, Dangers, Strategies.* Oxford: Oxford University Press Inc.

Caner, S., & Bhatti, F. (2020). A Conceptual Framework on Defining Businesses Strategy for Artificial Intelligence. *Contemporary Management Research* 16(3), 175–206. http://dx.doi.org/10.7903/cmr.19970.

Cheatham, B., Javanmardian, K. and Samandari, H. (2019). Confronting the Risks of Artificial Intelligence. *McKinsey Quarterly.* www.mckinsey.com/capabilities/quantumblack/our-insights/confronting-the-risks-of-artificial-intelligence. Accessed 11 October 2022.

David, L.O., Nwulu, N.I., Aigbavboa, C.O., and Adepoju, O.O. (2022). Integrating Fourth Industrial Revolution (4IR) Technologies into the Water, Energy & Food Nexus for Sustainable Security: A Bibliometric Analysis. *Journal of Cleaner Production* 363, 132522. https://doi.org/10.1016/j.jclepro.2022.132522.

David, L.O., Nwulu, N.I., Aigbavboa, C.O., and Adepoju, O.O. (2023). Resource Sustainability in the Water, Energy, and Food Nexus: Role of Technological Innovation. *Journal of Engineering, Design, and Technology.* https://doi.org/10.1108/JEDT-05-2023-0200.

Deloitte. (2017). *Managing Algorithmic Risks: Safeguarding the Use of Complex Algorithms and Machine Learning.* Stamford: Deloitte Risk and Financial Advisory.

Dogo, E., Nwulu, N.I., Twala, B., and Aigbavboa, C. (2019). A Survey of Machine Learning Methods Applied to Anomaly Detection on Drinking-Water Quality Data. *Urban Water Journal* 16(3), 235–248. https://doi.org/10.1080/1573062X.2019.1637002.

Dogo, E.M., Salami, A.F., Nwulu, N.I., and Aigbavboa, C.O. (2019). Blockchain and Internet of Things-Based Technologies for Intelligent Water Management System. In: Al-Turjman, F. (ed.) *Artificial Intelligence in IoT: Transactions on Computational Science and Computational Intelligence*. Cham: Springer. https://doi.org/10.1007/978-3-030-04110-6_7.

Dua, S., and Du, X. (2011). *Data Mining and Machine Learning in Cybersecurity*. Boca Raton: Auerbach Publications.

Federation of European Risk Management Associations. (2019). Artificial Intelligence Applied to Risk Management. www.eciia.eu/wpcontent/uploads/2019/11/FERMA-AI-applied-to-RM-FINAL.pdf.

Financial Stability Board. (2017). Artificial Intelligence and Machine Learning in Financial Services. www.fsb.org/wp-content/uploads/P011117.pdf. Last Accessed 17 August 2018.

Goertzel, B. (2006). *The Hidden Pattern*. Boca Raton: Brown Walker Press.

Howie, B. (2015). An Introduction to Artificial Intelligence. Archived from University of cincinattiUc.Edu.org. www.uc.edu/content/dam/uc/ce/docs/OLLI/Page%20Content/ARTIFICIAL%20INTELLIGENCEr.pdf.

ISO. (31000:2009). *Risk Management—Principles and Guidelines*. Geneva: International Organization for Standardization.

ISO/IEC. (2011). 27005 *Information Technology—Security Techniques—Information Security Risk Management*. Geneva: International Standard.

Kurzweil, R. (2005). *The Singularity is Near: When Humans Transcend Biology*. New York: Penguin.

McCarthy, J. (2007). *What is Artificial Intelligence?* Stanford: Computer Science Department, Stanford University, pp. 1–15.

Mckeown, P. (2021). What Are the Risks of Artificial Intelligence? www.auditboard.com/blog/what-are-risks-artificial-intelligence/.

McLean, S., Read, G.J.M., Thompson, J., Baber, C., Stanton, N.A., and Salmon, P.M. (2021). The Risks Associated with Artificial General Intelligence: A Systematic Review. *Journal of Experimental & Theoretical Artificial Intelligence* 35, 649–663. https://doi.org/10.1080/0952813X.2021.1964003.

Mumbai. (2014). Mumbai University, Introduction to Artificial Intelligence. https://mu.ac.in/wp-content/uploads/2014/04/Artificial-Intelligent-subjectM.SC_.IT-Part-2-.pdf.

National Institute of Standards and Technology. (2002). Risk Management Guide for Information Technology Systems, 1–55. https://nvlpubs.nist.gov/nistpubs/Legacy/SP/nistspecialpublication800-30.pdf

Ngai, E.W., Hu, Y., Wong, Y.H., Chen, Y., and Sun, X. (2011). The Application of Data Mining Techniques in Financial Fraud Detection: A Classification Framework and an Academic Review of Literature. *Decision Support Systems* 50(3), 559–569.

Olukanmi, S. Nelwamondo, F.V., and Nwulu, N. (2022). Leveraging Google Search Data and Artificial Intelligence Methods for Provincial-level Influenza Forecasting: A South African Case Study. *International Journal of Online and Biomedical Engineering (iJOE)* 18(11), 95–126.

Pathak, A. (2022). Top 14 In-Demand Skills Required for AI Professionals. https://geekflare.com/skills-required-for-ai-professionals/. Accessed 26 October 2022.

Quarterly, M. (2019). *Confronting the Risks of Artificial Intelligence*. McKinsey Global Institute, 8 September 2022.

Raaghieb, N., Yaeesh, Y., and Rashied, S. (2017). Risk Management. Archived from docplayer.net/4489777-The-university-of-adelaide-risk-management.

Salmon, P.M., Carden, T., and Hancock, P. (2021). Putting the Humanity into Inhuman Systems: How Human Factors and Ergonomics Can Be Used to Manage the Risks Associated with Artificial General Intelligence. *Human Factors and Ergonomics in Manufacturing & Service Industries* 31(2), 223–236.

Sanford, A., and Moosa, I. (2015). Operational Risk Modelling and Organizational Learning in Structured Finance Operations: A Bayesian Network Approach. *Journal of the Operational Research Society* 66(1), 86–115.

SBS CyberSecurity. (2018). Risk Assessment: Qualitative vs Quantitative. https://sbscyber.com/resources/risk-assessment-qualitative-vs-quantitative. Accessed 28 September 2022.

So, K. (2020). The Emergence of the Professional AI Risk Manager. *Towards Data Science*. https://towardsdatascience.com/the-emergence-of-the-professional-ai-risk-manager-d25d5b4822ef.

Sotala, K., and Yampolskiy, R.V. (2015). Responses to Catastrophic AGI Risk: A Survey. *Physica Scripta* 90(1), 1–33. Article 018001. https://doi.org/10.1088/0031-8949/90/1/018001.

Srinivas, K. (2019). *Process of Risk Management. In Book: Perspectives on Risk, Assessment and Management Paradigms*. IntechOpen. http://dx.doi.org/10.5772/intechopen.80804.

Stanton, N.A., Eriksson, A., Banks, V.A., and Hancock, P.A. (2020). Turing in the Driver's Seat: Can People Distinguish Between Automated and Manually Driven Vehicles? *Human Factors Ergonomics in Manufacturing Service Industries* 30(6), 418–425. https://doi.org/10.1002/hfm.20864.

Tegmark, M. (2017). *Being Human in the Age of Artificial Intelligence*. New York: Vintage Books.

Tom, B., Suchitra, N., Sulabh, S., and Alan, T. (2018). Centre for Regulatory Strategy EMEA. https://www2.deloitte.com/content/dam/Deloitte/nl/Documents/innovatie/deloitte-nl-innovate-lu-ai-and-risk-management.pdf.

Tutorials Point. (2015). Artificial Intelligence—Overview. www.tutorialspoint.com/artificial_intelligence/artificial_intelligence_overview.htm. Accessed 27 September 2022.

Vallor, S. (2017). *The Real Risks of Artificial Intelligence*. Swiss Re Centre for Global Dialogue. 8 September 2022.

Zigiene, G., Rybakovas, E., and Alzbutas, R. (2019). Artificial Intelligence Based Commercial Risk Management Framework for SMEs. *Sustainability* 11(450). http://dx.doi.org/10.3390/su11164501.

5 The Internet of Things

5.1 INTRODUCTION TO AND COMPONENTS OF THE INTERNET OF THINGS

Kevin Ashton first used the term "Internet of Things" in 1991 (Li et al. 2015). It is a cutting-edge innovation that has been included in several pieces of processing equipment, allowing access to crucial information, and enhancing people's quality of life (Villamil et al. 2020). The Internet of Things can be described as "a system that performs various types of functions, such as services involved in device modeling, device control, data publishing, data analysis, and device detection" (Ray 2018). Moreover, due to its promising potential, the Internet of Things has dramatically superseded other technologies (Nord et al. 2019). However, various authors and academia have defined the Internet of Things (IoT) based on their perspectives. In Sengupta et al. (2020), IoT is described as "a group of interconnected static or dynamic mobile objects such as devices equipped with communication, sensors, and actuator modules connected through the internet." However, Chen et al. (2017) opine that "IoT connects sensing devices to the internet to exchange information." Siboni et al. (2019) opined that "IoT is a global ecosystem of information and communication technologies aimed at connecting any type of object (thing), at any time, and in any place, to each other and the internet."

Meanwhile, Al-kadhim and Al-Raweshidy (2019) identified that IoT comprises many sensor nodes with limited processing, storage, and battery abilities. Internet of Things (IoT) devices uses IP-based connectivity and short-range communication technologies, which include Bluetooth, ZigBee, Z-Wave, and near-field communication (NFC), to communicate with each other as well as machine-to-machine (M2M), with people and their environment (Hassan et al. 2013). The concept of the Internet of Things, which can turn physical objects into artificially intelligent ones, has grown to be incredibly adaptable and popular in information technology (Onibonoje et al. 2019). However, it seeks to unite these real-world things through a tiered framework that gives us access to them and allows us to determine each real-world object's current state (Madakam et al. 2015; Adeleke et al. 2023). IoT also intends to build a reliable, self-sufficient, and secure system for data exchange between connected devices and real-world applications. IoT also enables machine-to-machine communication and the incorporation of intelligence into gadgets. Based on this attribute, data may be analyzed quickly, allowing the system to make real-time decisions without requiring human participation (Khan et al. 2012; David et al. 2022; Adepoju et al. 2022; David et al. 2023). IoT solutions can be used in various fields, so they can be grouped into different

DOI: 10.1201/9781003522102-7

application domains in various ways. According to Atzori et al. (2010), the following are the domains for IoT applications:

Futuristic domain: The IoT application space of the future includes advanced gaming environments, city information models, robot taxis, and so forth.
Smart environment domain: This domain includes industrial facilities, modern museums and fitness centers, cozy workplaces, houses, and so forth.
Healthcare domain: Data gathering, authentication, identification, tracking, sensing, and other IoT applications are used in this field.
Personal and social domain: This domain's applications include historical search, social networking, loss and theft prevention, and so forth.
Transportation and logistics domain: Applications of this domain include mobile ticketing, logistics, aided driving, and augmented mapping.

In addition, Gubbi et al. (2013) divided IoT applications into four application domains:

1. **The personal and home domain:** These domain applications include prevention of privacy theft/intrusion, social networking, and so forth.
2. **The enterprise domain:** These applications include product recommendation, data analysis, business optimization, and so forth.
3. **Utilities:** This domain's application includes a smart grid, smart meter, and so forth.
4. **Mobile:** Applications of this domain include mobile biometric authentication, GPS, and so forth.

However, Sundmaeker et al. (2010) also divided IoT applications into the following three application domains in their study:

Environment domain: This domain application involves weather forecasts, agriculture, and so forth.
Industrial domain: The industrial domain includes utilities, enterprise, health care, and other industries.
Social domain: This domain application includes social networking, and so forth.

Figure 5.1 clearly depicts the definitions of IoT integrating all the possible aspects of communications and information-sharing processes between various objects.

In the deployment of the Internet of Things (IoT), there are four essential building blocks: things, gateways, network infrastructure (NI), and cloud infrastructure (CI) (Ahmed 2017). Things in the Internet of Things (IoT) components, including sensors and actuators, guarantee information collection and communication without human involvement. On the other side, gateways serve as a connecting point between entities, such as network infrastructure or cloud infrastructure. The network infrastructure (NI) block is intended to offer management for safe and efficient information transport. These network infrastructure components could be repeaters, gateways, aggregators, routers, and so forth.

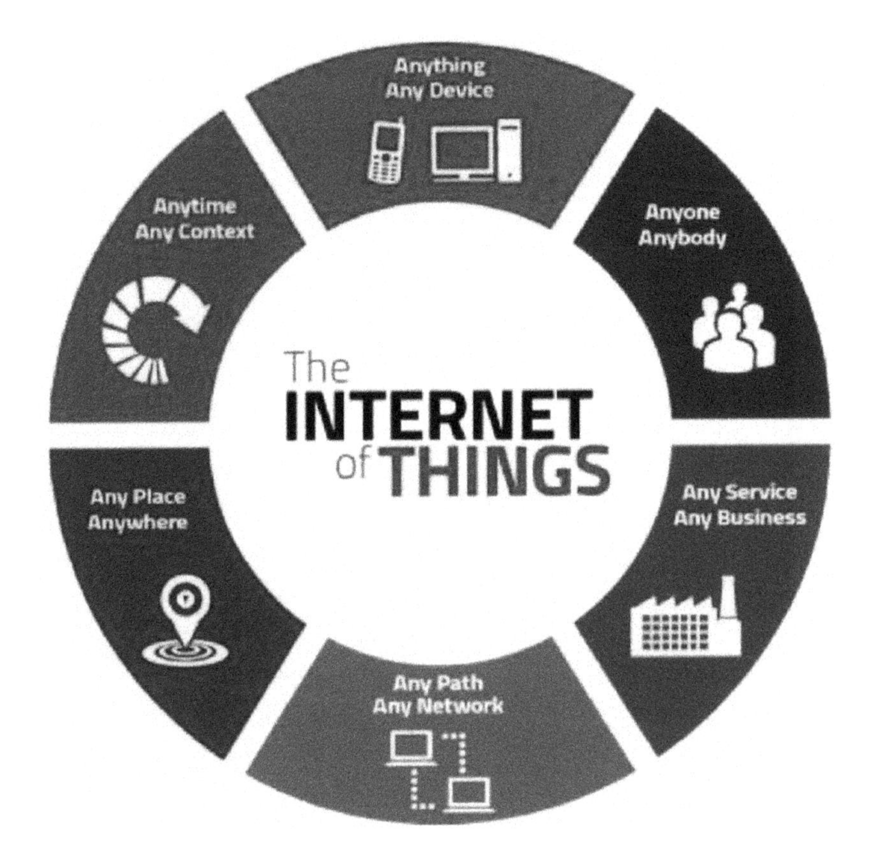

FIGURE 5.1 Overview of the Internet of Things (Kumar and Mallick 2018, p. 110)

On the other hand, IoT may use enhanced computational capabilities thanks to cloud infrastructure like virtualized servers (VS) and data storage units (DSU). However, the four building blocks described here are then divided into three crucial elements to thoroughly understand IoT's process flow and implementation. The three primary IoT components are things with networked sensors and actuators (TNSA), raw information and processed data stores (RI-PD-S), and analytical and computing engines. The three IoT components are listed in Table 5.1, illustrating each one's attributes (Kumar and Mallick 2018).

The interactions between the three elements in Table 5.1 are shown in Figure 5.2. The following three elements interact: The first component, things with networked sensors and actuators, gathers data following user requirements. The second component, which is utilized to store the sensed data, is called raw information and processed data storage. Depending on the sort of data being sensed, different components, such as text, data, videos, and photos, are sent to the second component to enable this interaction. The final component, analytical computing engines (ACE), is in charge of logically assessing the information

TABLE 5.1
Features of IoT components

IoT components	Features
Things with networked sensors and actuators	Gathering data from objects or things that were concentrated per the relevant application area.
Raw Information and processed data stores	stores the gathered data in a variety of formats, including data, text, videos, photos, models, and so forth.
Analytical and computing engines	They facilitate interactions between people and machines and enable feedback tailored to human needs. Additionally, they permit analysis based on the computing model picked based on the needs.

Source: Kumar and Mallick (2018).

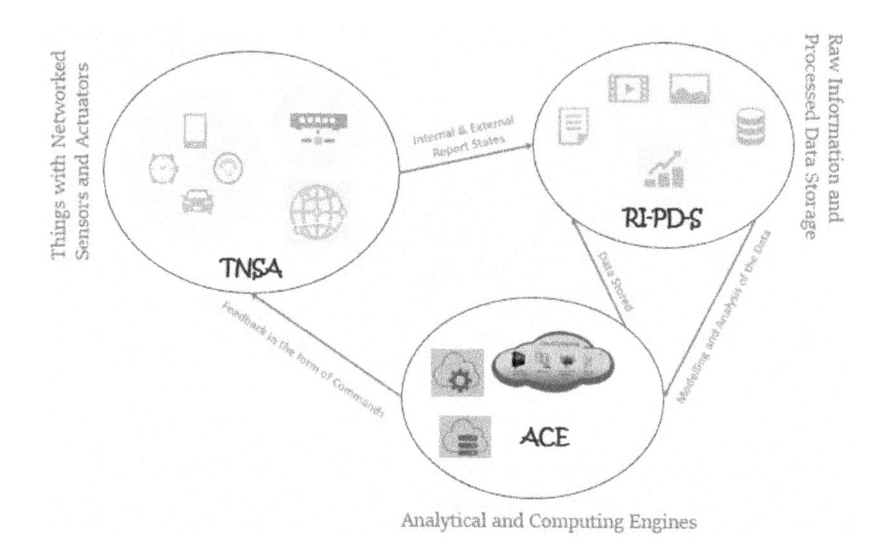

FIGURE 5.2 IoT component interactions (Kumar and Mallick 2018, p. 112)

that has been saved. Multiple iterative techniques are also employed for logical analysis to get the user's desired outcomes. The third element guarantees the viability of cloud-based, server-based, and human-to-machine learning (Ram 2015).

Furthermore, according to Zanella et al. (2014), because of the enormous networks and connected devices, creating a universal architecture for the IoT is highly laborious. Figure 5.3 illustrates how IoT design can be separated into three layers: the network layer, the application layer, and the perception or recognition layer (Zhang et al. 2012; Kumar and Patel 2014; Zhao and Ge 2013). Additionally, the Internet of Things offers intriguing remedies to most of the workforce's issues. The method of arriving at a solution is based on how the communication devices and information technology components were combined with the best hardware and software convergence. In the Internet of Things, the hardware and software components interact and cooperate with the owner's learning outcome-based objectives. In short, the software-defined hardware system aids in the transformation of raw data into processed data, from which advanced computing tools coupled with IoT systems perform data storage, retrieval, and analysis. The communication systems aid in supplying the channels for communication and enable protocols amongst the items or devices that make up each IoT component. Only when a strong IoT architecture layer is created will information technology and communication technology merge in the best, fastest, most reliable, and most secure manner. These architecture layers would change depending on the needs and duties to be handled. However, several researchers have put forth various architectures, including the ones shown later. This section provides a summary of the generally recognized and acknowledged IoT architecture. One of the first and most fundamental IoT architectures described is the three-layer architecture depicted in Figure 5.3. It is incredibly practical and simple to use. The perception layer, network layer, and application layer are the three layers that make up the architecture (Mashal et al. 2015; Said and Masud 2013; Wu et al. 2010; Pallavi and Smruti 2017). The role and functional components of the three-layer IoT architecture are shown in Table 5.2.

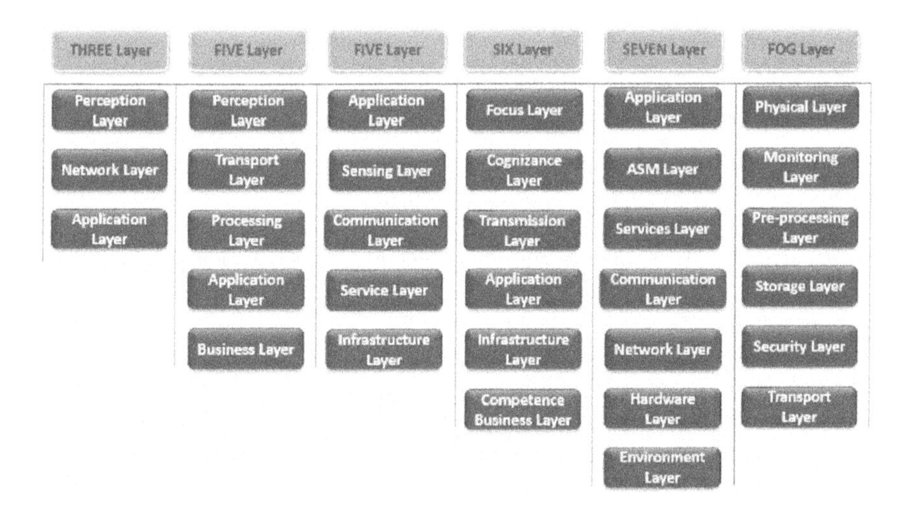

FIGURE 5.3 IoT architecture layers (Kumar and Mallick 2018, p. 113)

TABLE 5.2
Role and functional features of three-layer IoT architecture

IoT Architecture Layers	Role and Functional Features
Perception Layer	The feature of this layer would be the sensing capability; it collects and gathers information about the environment where smart objects are available.
Network Layer	The layer's feature enables the transmission and processing of information with Internet connectivity of the various devices.
Application Layer	Its main feature is to deliver a specific service to the user based on the application type.

Source: Kumar and Mallick (2018).

5.2 RISKS ASSOCIATED WITH THE INTERNET OF THINGS

There are still certain risks involved with the Internet of Things despite the premise of ensuring that devices are always connected to the Internet via Internet Protocol (IP) and its promising potential. Majorly, these risks are Privacy and security. This is because establishing security and privacy in the IoT is highly difficult, especially for devices with limited resources like low computational power or low energy (Samaila et al. 2018). According to a study by Banafa (2014), IoT has three crucial obstacles that take a broader view. These consist of:

a. Technological challenges
b. Societal problems/risks
c. Business challenges/risks

5.2.1 Technological Challenges

IoT implementation components are constructed using a variety of protocols and technologies. Although many different protocols and technologies exist, the components still have poor designs and configuration problems. Sundmaeker et al. (2010) claim that five key variables are to blame for these technological issues in their study.

1. **Connectivity:** One of the main obstacles to the future of the Internet of Things is the connectivity of numerous devices on a network. However, this is because connecting several devices will interfere with the communication technologies and protocols that are already in use. For instance, a single network could aid in the interconnection of billions of devices, but this is subject to online data storage, which unavailability may lead to network collapse, which may lead to several damages to applications in several fields. A new server architecture is on

the horizon, which aids the connection, authentication, and access of devices to several networks. This is also subject to the issues of inconsistency in the connection process, as it is mostly dependent online (Vermesan et al. 2011).

2. **Standards:** This entails the non-uniformity of standards for different data arrays used by different IoT applications. Hence, standards in network protocols, communication protocols, and data aggregation facilitate the handling, processing, and storing of data obtained from the sensors, which, because of this aggregation and non-uniformity, data amount, breadth, and frequency all rise and varies (Archana and Vinodhini 2017). For instance, the standards of storage of both structured and unstructured data in different databases raises issues of concern, especially on the different levels of IoT applications.
3. **Security:** IoT security has been a top priority. Additionally, security problems in IoT applications are experienced by the public and private sectors worldwide. Adding new hubs to the IoT architecture and the Internet creates opportunities for hackers to penetrate the system and eavesdrop, which are now evidence in hacking laptops, security gates, smart fridges, CCTVs, camera, drug infusion pumps, intelligent assault rifles, and thermostats. In this regard, the malware used by these attackers can detect an endless number of Internet of Things (IoT) devices linked to the network, such as smart home appliances and closed-circuit televisions, and then turn those devices against their servers. Furthermore, new security-related difficulties will arise given how IoT has been incorporated into every aspect of our lives (Sundmaeker et al. 2010).
4. **Intelligent analysis and actions:** Data from IoT devices or data processed into IoT applications are analyzed, especially from artificial intelligence and cloud computing models. However, there exist the challenge of inaccuracy of data, which leads to incorrect algorithms, and leads to consequential business loss. Moreover, less expensive devices, more functional devices, machines that "influence" human behavior using behavioral science, deep learning tools, machines acting in unexpected situations, information security, and privacy, and device interoperability are some of the factors that lead to intelligent actions being incorporated in IoT (Theoleyre and Pang 2013; Adeleke, Nwulu, and Adebo 2023). These factors aid the generation of different structured and unstructured data, which are inaccurate due to several human and environmental issues, thus becoming a risk for IoT applications.
5. **Compatibility and longevity:** Deploying additional software and hardware while connecting IoT would be highly challenging due to the standardization rivalry between many technologies (Archana and Vinodhini 2017). Therefore, this raises issues of compatibility of different servers with different standards with different IoT applications based on their needs. Thus, leading to short-lived applications of many IoT devices, which acts as a risk to their longevity.

5.2.2 Societal Problems/Risks

These kinds of risks explain the risks associated with a paradigm change in society involving producers, regulators, and clients. A thorough understanding of the implementation of IoT from regulators' and clients' points of view is very challenging due to the following reasons:

- There are consistent changes in consumption patterns and customer requirements.
- Drastic growth and development of new devices, which are complementary and most often competing in nature.
- Growing IoT technology.
- Consumer confidence is being tampered with (Coetzee and Eksteen 2011).

5.2.3 Business Challenges/Risks

IoT technologies have been adopted in generating multiple streams of income for different business and organizations, which has reduced stress on communication infrastructure but brings to limelight that an IoT system without a foolproof plan of action would be a catastrophe to business models and strategy (Banafa 2014). Also, the utilization of IoT technology has led to more costs in business occasioned by IoT system maintenance costs and the costs arising from security issues. Expenditures are essential to any technology, and installation, infrastructure, and labor costs for the implementation phase are included. These might be a barrier to developing an IoT application for business (Villamil et al. 2020).

Moreover, many firms such as finance, supply chain, and healthcare are impacted by IoT attacks. Hence, examples of the different types of IoT risks are given in the following (Kandasamy et al. 2020) and depicted in Figure 5.4.

Ethical IoT: This risk encompasses the potential unforeseen repercussions arising from unethical practices and actions involving IoT devices, as Zhou (2016) highlighted. This risk category emphasizes the importance of ethical considerations in the rapidly evolving landscape of Internet of Things technologies, where the misuse or unintended consequences of connected devices may have ethical implications that extend beyond their initial design and deployment.

Security and privacy IoT risk: This entails system vulnerabilities, where malicious actors exploit weaknesses to gain unauthorized access to valuable assets, intending to inflict harm. Antonakakis et al. (2017) emphasize the critical nature of safeguarding security and privacy in the IoT ecosystem. This risk underscores the need for robust cyber security measures to protect sensitive information and maintain the integrity of interconnected systems, as the consequences of security breaches can be far-reaching.

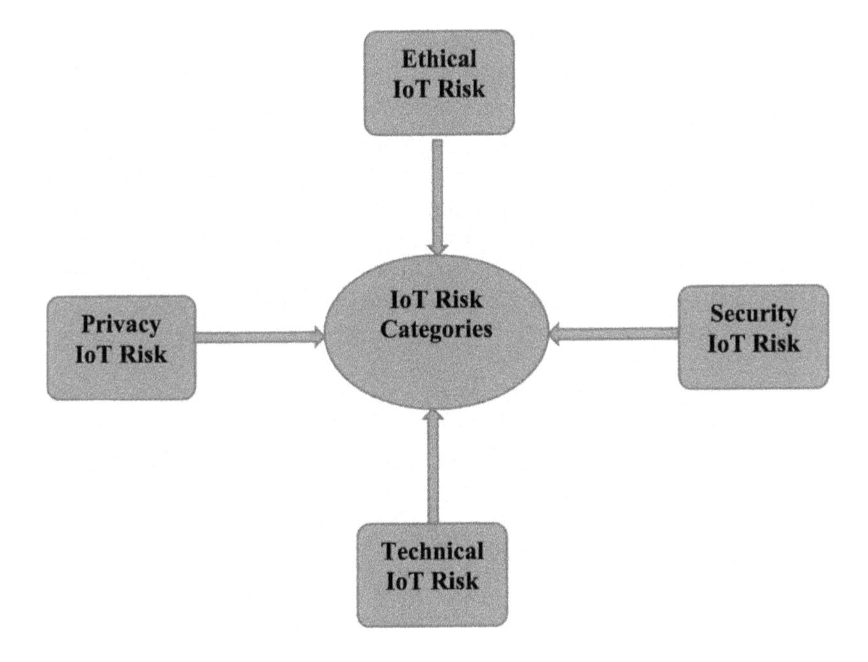

FIGURE 5.4 IoT risk categories (Kandasamy et al. 2020, p. 3)

Technical IoT risk: This revolves around the challenges posed by hardware or software failures within IoT devices, often attributed to inadequate design, evaluation, or implementation processes. This risk highlights the significance of technical excellence in developing and deploying IoT technologies. Addressing technical risks involves ensuring the reliability and resilience of hardware components and software systems, acknowledging that failures in these areas can have profound implications for the functionality and performance of IoT devices.

5.3 RISK MANAGEMENT TECHNIQUES IN THE INTERNET OF THINGS

The Internet of Things (IoT) could also be used more than ever to manage risks thanks to the adoption of suitable systems, safeguards, and security standards. This was evidenced in the use of IoT technologies in monitoring and controlling physical and social distance during the COVID-19 pandemic in 2020. However, it offers intriguing potential with substantial advantages for the risk profession and the public (Boler 2020).

However, as stated earlier, IoT risks usually emanate from privacy and security issues, mostly penetrated via the networks used in the IoT application. However, most network protocols are not created with security in mind, so most of them have separate network security software and applications. Additionally, the accessibility of these network elements could turn into a way for attackers to penetrate the network. Therefore, managing this network security issues which is a risk to IoT applications, it may be necessary to partition the network's devices into two groups: trustworthy devices and untrusted devices. The first component is categorized as having a "root-of-trust," which could be something as straightforward as an attestation key provided by the manufacturer (Schiffman et al. 2011).

The trusted devices use safe key storage, cryptographic operations, and other secure communication techniques. Attestation techniques are used to determine whether the untrusted devices are sufficiently secure to join the group of trusted devices. Because it is challenging to categorize those categories according to kind, use, and maker, the distinction between the trusted devices group and the untrusted ones is essentially unnoticeable. The attestation procedure, meanwhile, is dynamic and frequently modifies itself in response to network threats. Root-of-trust in IoT refers to a group of device features that are believed to be reliable and unlikely ever to be compromised (Abera et al. 2016).

A root of trust is, for instance, the device's secure booting feature. Another illustration is the attestation functionality, which uses cryptographic techniques to demonstrate the reliability of chains. An IoT device may have several trust-based foundations. Any IoT framework's ecosystem for establishing trustworthiness comprises several building elements, as depicted in Figure 5.5. The trusted

FIGURE 5.5 Building blocks of IoT trustworthiness (Showail 2021, p. 3)

execution environment (TEE), where trusted application programs and security risk minimization are executed, is the first building block. Additionally, it oversees dividing active processes from those running on the same hardware.

The second component is the secure communication channel, which protects the secrecy and integrity of the data traveling between devices in the network using conventional encryption algorithms. The third part of the system is the authentication procedure, which includes both the keys—symmetric or asymmetric—and the key distribution and authentication protocol itself. The attestation procedure, which consists of the verification logic and the attestation key supplied by the manufacturer, is the fourth element. The devices must then be capable of safely storing all the keys and information gathered from the sensors. The capacity to compile all pertinent contextual data, including time and location, is the final capability (Showail 2021).

Most notably, the end-to-end notion is essential for communication network security. The key elements in the end-to-end security journey of IoT device communication are depicted in Figure 5.6. Before being transmitted to the gateway, the sensor data is first encrypted. Sometimes, when the encryption procedure is complete, these data are stored locally. The gateway then decrypts the data, runs some data analytics, re-encrypts the data, and permits cloud sharing. The cloud will need to interpret the data after receiving it and perform some analytics before the data can be saved in the database. IoT frameworks have also been developed to increase networks' portability and use. The intricacy of the network topology and

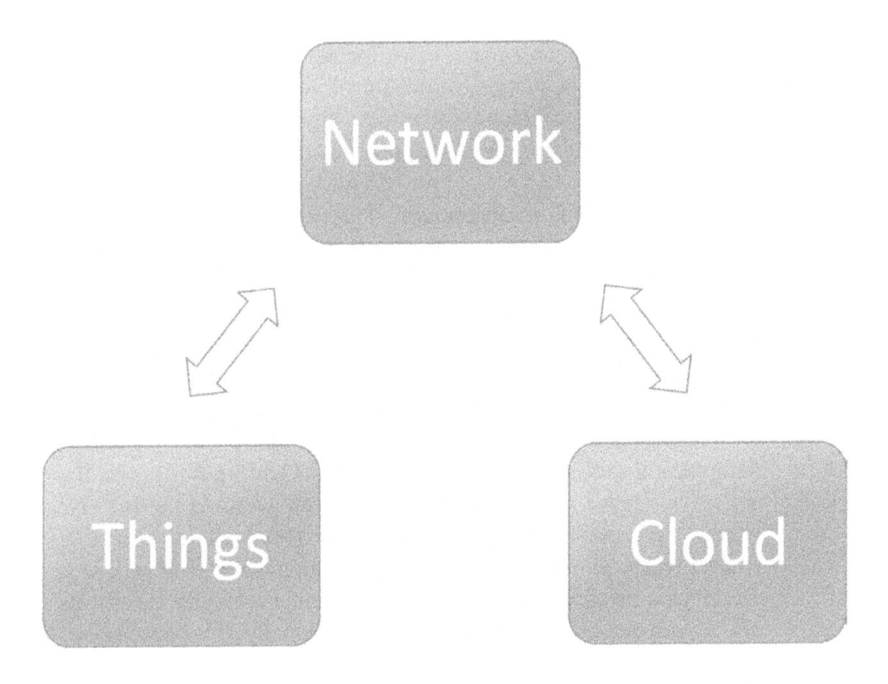

FIGURE 5.6 Components of end-to-end security in IoT (Showail 2021, p. 5)

architecture is concealed by this framework, which acts as a "black box" technology (Showail 2021).

Furthermore, in managing the risks of the Internet of Things, there are different frameworks utilized, which are explained thus:

The ISO risks management framework for IoT: According to Tiganoaia et al. (2019), the ISO 31000:2009 internal standards are necessary for managing risk, given the principles, framework, and processes involved. This is also corroborated in the research of Ben-Daya and Akram (2013). The risk management framework includes the following:

i. **Mandate and commitment:** To effectively deploy a risk management strategy, tailored to the unique needs of various IoT applications, full commitment from the leadership and management of the parent organization is essential. This entails engaging with diverse stakeholders, allocating adequate resources, establishing clear policy directives, delegating responsibilities effectively, and ensuring strict adherence to the principles of the chosen risk management strategies as needed for an IoT application or product.
ii. **Design of framework for managing risk:** When designing a risk management strategy or protocol for an IoT product, it's essential to take into account various elements. These include internal and external factors, industry trends, contractual obligations, the organization's historical product management perspectives, and future strategic initiatives. This requires thorough stakeholder engagement, review of existing organizational policies, and regular interaction between employees possessing technical and soft skills. Additionally, it involves evaluating directives, standards, and regulations from relevant regulatory bodies applicable to the deployment of the IoT application.
iii. **Implementing risk management:** This involves implementing a strategy or technique that has been unanimously selected and evaluated by the relevant teams for the IoT application. This phase encompasses applying the risk management policy, conducting the risk management process, meeting compliance requirements, and ensuring sound decision-making. However, during this implementation phase, it is advisable to employ the agile project management approach, which involves an iterative review of processes after each cycle.
iv. **Monitoring and review of the framework:** During this phase, the activities revolve around assessing the effectiveness of the IoT risk strategy in meeting its defined objectives. This includes analyzing risk reports to pinpoint any deviations, gauging the level of progress, identifying areas requiring adjustments, and evaluating stakeholders' alignment with the risk strategy. Additionally, this phase serves to uncover any weaknesses within the team responsible for implementing the IoT application and product.

v. **Continual improvement of the framework:** This stage involves implementing corrective actions based on the insights gained from previous steps to enhance the risk strategy for IoT products. Continuous improvement of the framework ensures its usability and iterative development, encompassing the risk management framework, plan, and policy across all levels. However, considering the dynamic nature of IoT, while enhancing performance, it's imperative to communicate the need for product improvement to relevant stakeholders, along with clear timelines for implementation and restoration of services.

Furthermore, from the established IoT risk management framework, the process for risk management for Internet of Things (IoT) products adapted from the ISO framework are now briefly explained.

Specifying and establishing context: This process involves defining the context of the IoT product, which includes outlining its scope of operation, intended purposes, application areas, and expected outcomes. It also entails specifying the operational framework for the chosen risk strategy, determining required resources, and deployment areas, and assessing both internal and external deployment environments. Additionally, criteria indicating uncertainties during deployment are identified to ensure comprehensive risk assessment and management.

Risk assessment: As outlined in the ISO 31000 standard, the risk assessment process comprises three key stages: risk identification, risk analysis, and risk evaluation. During risk identification, potential risks,

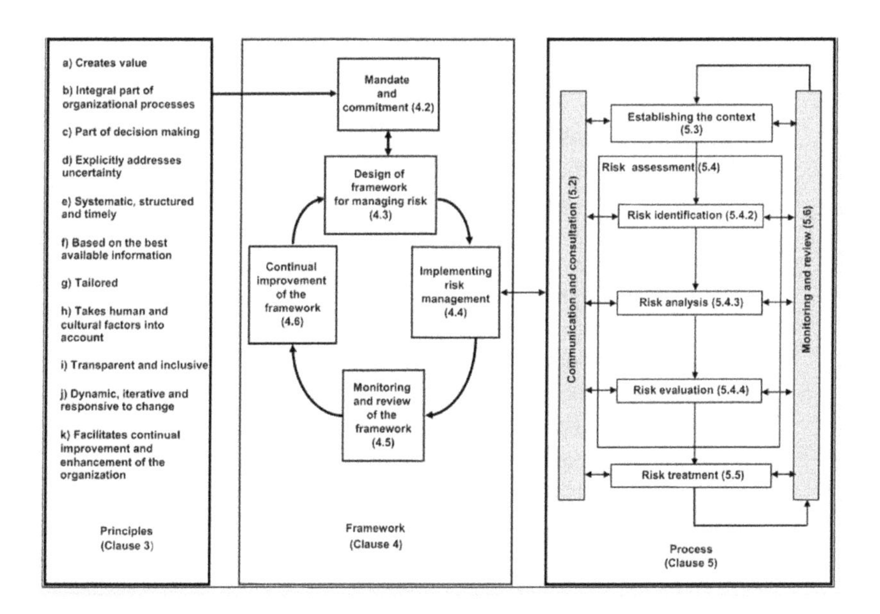

FIGURE 5.7 ISO risk framework Tiganoaia et al. (2019, p. 2)

their origins, and the potential negative impacts they may have are identified. For IoT products or applications, leveraging expert opinions, user records, and sensor data from Internet-based sources is crucial in this phase. Subsequently, in the risk analysis phase, the causes and sources of risks are examined. The primary objective here is to assess the quantity of identified risks, their potential consequences, complexity, alignment with objectives, and associated costs. The third stage, risk evaluation, involves developing and prioritizing the results obtained from risk analysis for further action. This phase helps determine the overall consequences of risks, as well as the organization's and stakeholders' risk tolerance levels and decision-making capabilities regarding identified risks for IoT products or applications.

Risk treatment: Once risks have been assessed, the subsequent step involves treating them, which includes selecting suitable approaches to mitigate the risks. These approaches may entail avoidance, acceptance, elimination of risk sources, modification of likelihood or impact, risk sharing, or retention. It's crucial to develop a comprehensive risk treatment plan that considers all aspects of each option, including economic implications, human resources, technological factors, stakeholder viewpoints, cost-benefit analysis, implementation timelines, effects on resources, and available contingencies. This ensures a thorough evaluation of each option's multiplier effects and enables informed decision-making to effectively manage and address the identified risks.

Communication and consultation: This phase involves extensive communication both within and among stakeholders involved in the IoT products and applications. It entails engaging and consulting with all pertinent parties to ensure that the products align with their expectations and requirements and that the adopted risk strategy is widely acknowledged and embraced. It's essential for communication to be regular and timely, addressing emerging needs, and utilizing mediums that are universally accepted. Given the dynamic nature of the Internet of Things, real-time communication is imperative to promptly detect any deviations or changes that may occur. This ensures that all stakeholders remain informed and involved throughout the process.

Monitoring and review: This process entails monitoring and evaluating the performance of all elements and overall effectiveness throughout the risk management process, thereby necessitating conducting regular assessments based on current situations of usage. This involves documenting the operational dynamics of IoT products and establishing a feedback mechanism. Additionally, it ensures adherence to quality standards and specifications for IoT products. Moreover, as suggested earlier employing the iterative approach of agile project management ensures consistent, timely monitoring of the entire lifecycle of IoT products.

COBIT5 Framework for IoT risk management: This framework is an IT standard that covers the IoT risks in terms of data and application, physical environment, change management, third-party suppliers and vendors, security and privacy, and infrastructure (Latifi and Zarrabi 2017). It's an acronym for "control objectives for information and related technology." It offers insights into dealing with IoT hazards more effectively and efficiently while reducing the effort, expense, and time required to do so. Additionally, the influence factors in managing IoT risk in a business were clearly described in this model. The COBIT5 model is shown in Figure 5.8 (Latifi and Zarrabi 2017). As stated by the author, the framework ensures the security and reliability of data, enhances and measures data performances of IoT devices, mitigates negative risks, reduces complexity in managing IoT risks, and reduces business interruptions.

ISO 31010: A collection of risk assessment techniques: This method offers recommendations for the best choice and use of systematic risk assessment methodologies. However, it summarizes how to put the ISO 31000 criteria into practice. The complexity of the problem and the method of analysis, the application within the many process activities that determine the methodology type, and other factors set this risk assessment approach apart from others. Additionally, ISO 31010 has advantages over other procedures in that it covers several risk assessment methodologies and offers appropriate information for which risk process activities can yield quantifiable outcomes (ISO 2009).

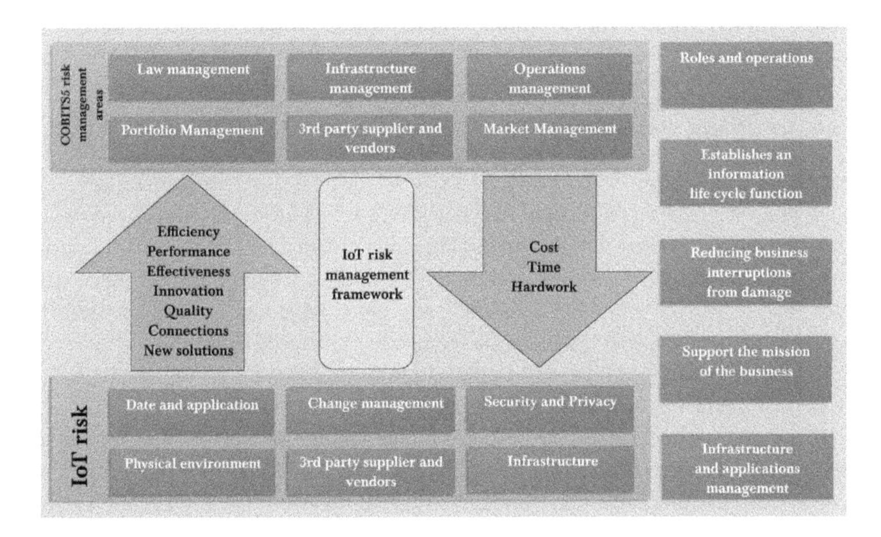

FIGURE 5.8 COBIT5 framework for IoT risk management (Latifi and Zarrabi 2017, p. 42)

OCTAVE for IoT: OCTAVE is a risk assessment tool used in smart homes. This is because it may address both physical and cyber security (Ali and Awad 2018). However, it aids in identifying various IoT-based smart home security flaws, highlights hazards to residents, and suggests ways to reduce the risks that have been found. On the other hand, OCTAVE categorizes the impact of recovery using a standard questionnaire but does not estimate risk. It goes through four stages, which are described in the following:

- Establish driver's phase: The strategy for measuring risks is developed in this phase.
- Profile assets phase: It identifies security requirements and establishes limits for assets.
- Identify threats phase: This phase identifies security threats from the assets where the information asset is stored, transported, or processed.
- Risk mitigation phase: This phase determines and executes a risk mitigation strategy for the identified assets.

TARA for IoT: TARA (time-alignment-value-based random access) is a framework for predicting the most significant exposures that may affect an IoT application. The three main benefits of TARA are as follows. It reduces potential assaults to a manageable list of likely assaults. It raises the standard of risk and control assessments and informs the organization of hazards and suggestions. It can improve results, lessen the overall effort required for risk analysis, and aid decision-making. It was created for Intel (Rlarge)'s extremely valuable, diverse environment in response to the necessity to evaluate the security risks of a complex, rapidly changing threat scenario. TARA does not emphasize vulnerability protection or quantify the impact of risks (Kandasamy et al. 2020).

5.4 RISK MANAGEMENT SKILLS FOR THE INTERNET OF THINGS

The Internet of Things (IoT) involves linked devices that frequently generate enormous amounts and types of data that are used, delivered to, or stored in various parts of an organization's IT infrastructure. This creates a cascading effect across the risk environment, including compliance, business resiliency, third-party risk, and cyber security. Companies will likely need to change their security strategy to handle IoT threats properly, whether it's due to the increased demand for discovery, identification, and classification of new endpoints, additional compliance checks, or improvements to authentication (Fontaine 2020).

In addition to controlling how data sources behave and restricting interested parties' access to resources, it's crucial to assess the platform's reliability that controls the supplied data and services. The foundation of the Internet of Things (IoT) is the connection between physical devices and any objects that may include gadgets using network architecture to enable physical resources such as sensors, actuators, or actuators. The difficulties or dangers that could appear have financial insecurity, legal liability, strategic management errors, mishaps, and natural disasters, to name just a few (Salamai 2021). All these can be managed via a competent skill set for IoT applications. There is a need for a combination of IoT risk management knowledge and skills in optimizing the knowledge. Furthermore, the risk management skills for the Internet of Things can be soft or hard. The soft skills include the following:

Analytical risk assessment skills: Risk managers must possess analytical skills to predict the risk in the Internet of Things. This skill includes collecting and analyzing data to make strategic decisions (Rapid 2021).

Problem-solving skill: Once potential risks are detected in the Internet of Things, one must be able to look into the application to proffer solutions to those risks (Rapid 2021). This requires a curious mind to understand the different segments of an IoT application and potential and similar risks involved in the application. This will enable quick detection of problems and provide solutions on one's feet.

Strategic thinking: Considering the complexity of the Internet of Things, strategic thinking is essential. Strategic thinking helps to look at every possible area where risks could emanate in the Internet of Things (Rapid 2021). Strategic thinking depends on the knowledge of strategic management plan of an organization and knowing the role of an IoT application.

Financial knowledge skills: The ability to prepare for unknown events is crucial in the Internet of Things. As every business has some element of financial risk, the Internet of Things is not exempted. Hence, financial knowledge skills are needed because the usual IoT application to business and organization processes involves money collection or financial traceability (Rapid 2021).

Furthermore, the technical skills required to learn risk management in the Internet of Things include and are not limited to the following:

Programming languages: To become an IoT developer, one must be conversant with the relevant languages. Currently, JavaScript and Java are the most popular languages used for programming IoT devices. Also, other languages, such as Python, and so forth are utilized too (Misal 2019).

Hardware skills: Aside from having programming skills, an IoT developer must understand how the hardware works. This is because providing solutions to software problems requires prior knowledge of hardware devices (Misal 2019).

Server languages: As IoT devices use data, server languages are required to develop servers for receiving and storing data from IoT devices. These server languages are used to manage connective devices like Raspberry Pi and Arduino. Some of the popular server languages used in IoT are Node.js, PHP, and ASP.NET (Misal 2019).

Artificial intelligence and machine learning: AI and ML are two crucial technical skills for an IoT developer. They help with the automation of IoT devices. However, machine learning skills help an IoT developer to analyze and predict patterns using data (Misal 2019).

UI-centric approach: UI (user interface) is essential when developing an IoT platform; hence, a skill that helps build interactive interfaces is crucial for IoT developers (Misal 2019).

5.5 SUMMARY

The building blocks and components of IoT were discussed in this chapter in conjunction with the application domains of IoT, such as the transportation and logistics, healthcare, and social domains. The chapter also covers the different risks associated with the Internet of Things. The risk management techniques for the Internet of Things were discussed. These involve specifying and establishing the context, risk assessment, risk treatment, communication and consultation, monitoring, and review. Also, the COBIT5 framework for IoT risk management, the conceptual model risk management techniques, ISO-31010-A collection of risk assessment techniques, OCTAVE for IoT, and TARA for IoT were discussed in this chapter. Finally, the risk management skills in the Internet of Things were divided into soft and technical skills. The soft skills discussed include analytical risk assessment, problem-solving, strategic thinking, and financial knowledge. At the same time, the technical skills include programming languages, hardware skills, server languages, artificial intelligence, machine learning, and UI-centric approach.

REFERENCES

Abera, T., Asokan, N., Davi, L., Koushanfar, F., Paverd, A., Sadeghi, A.R., and Tsudik, G. (2016). Things, Trouble, Trust: On Building Trust in IoT Systems. *Proceedings of the 53rd Annual Design Automation Conference*, pp. 1–6.

Adeleke, I., Nwulu, N., and Adebo, O.A. (2023). Internet of Things (IoT) in the Food Fermentation process: A Bibliometric Review. *Journal of Food Process Engineering* 46, e14321. https://doi.org/10.1111/jfpe.14321.

Adepoju, O., Aigbavboa, C., Nwulu, N., and Olaiya, M. (2022). *Reskilling Human Resources for Construction 4.0. Implications for Industry, Academia, and Government.* Springer. https://doi.org/10.1007/978-3-030-85973-2.

Ahmed, B. (2017). IoT and Blockchain Convergence: Benefits and Challenges. https://iot.iee.org/newsletter/january-2017/iot-and-blockchain-convergence-benefits-and-challenges.html. Accessed 12 August 2022.

Ali, B., and Awad, A.I. (2018). Cyber and Physical Security Vulnerability Assessment for IoT-Based Smart Homes. *Journal of Sensors* 18, 817. https://doi.org/10.3390/s18030817.

Al-Kadhim, H.M., and Al-Raweshidy, H.S. (2019). Energy Efficient and Reliable Transport of Data in Cloud-Based IoT. *IEEE Access* 7, 64641–64650.

Antonakakis, M., April, T., Bailey, M., Bernhard, M., Bursztein, E., Cochran, J., Durumeric, Z., Halderman, J.A., Invernizzi, L., Kallitsis, M., and Kumar, D. (2017). Understanding the Mirai Botnet. *Proceedings of 26th USENIX Security Symposium.* https://www.usenix.org/system/files/conference/usenixsecurity17/sec17-antonakakis.pdf

Archana, V., and Vinodhini, S. (2017). Fundamentals and Applications of IoT. *IOSR Journal of Computer Engineering (IOSR-JCE)* 6(5), 20–23. https://www.iosrjournals.org/iosr-jce/papers/Conf.17031-2017/Volume-6/5.%2020-23.pdf?id=7557.

Atzori, L., Iera, A., and Morabito, G. (2010). The Internet of Things: A Survey. *Computer Networks* 54(15), 2787–2805.

Banafa, A. (2014). *IoT Standardization and Implementation Challenges*. IEEE. Org Newsletter.

Ben-Daya, M., and Akram, M. (2013). Third-Party Logistics Risk Management. *Proceedings of 2013 International Conference on Industrial Engineering and Systems Management (IESM)*, Agdal, Morocco, pp. 1–10. https://ieeexplore.ieee.org/stamp/stamp.jsp?tp=&arnumber=6761469.

Boler, J. (2020). The Internet of Things (IoT) and the Management of Risk. www.theauditoronline.com/the-internet-of-things-iot-and-the-management-of-risk/. Accessed 13 August 2022.

Chen, L., Thombre, S., Jarvinen, K., Lohan, E.S., Alen-Savikko, A., Leppakoski, H., Bhuiyan, M.Z.H., Bu-Pasha, S., Ferrara, G.N., Honkala, S., Lindqvist, J., Ruotsalainen, L., Korpisaari, P., and Kuusniemi, H. (2017). Robustness, Security and Privacy in Location-Based Services for Future IoT: A Survey. *IEEE Access* 5, 8956–8977.

Coetzee, L., and Eksteen, J. (2011). The Internet of Things—Promise for the Future? An Introduction. *IST-Africa Conference Proceedings*, IEEE, pp. 1–9.

David, L.O., Nwulu, N.I., Aigbavboa, C.O., and Adepoju, O.O. (2022). Integrating Fourth Industrial Revolution (4IR) Technologies into the Water, Energy & Food Nexus for Sustainable Security: A Bibliometric Analysis. *Journal of Cleaner Production* 363, 132522. https://doi.org/10.1016/j.jclepro.2022.132522.

David, L.O., Nwulu, N.I., Aigbavboa, C.O., and Adepoju, O.O. (2023). Resource Sustainability in the Water, Energy, and Food Nexus: Role of Technological Innovation. *Journal of Engineering, Design, and Technology*. https://doi.org/10.1108/JEDT-05-2023-0200.

Fontaine, A. (2020). The IoT Domino Effect: Five Steps to Manage IoT Risk. www.information-age.com/iot-domino-effect-five-steps-manage-iot-risk-123490841/. Accessed 13 August 2022.

Gubbi, J., Buyya, R., Marusic, S., and Palaniswami, M. (2013). Internet of Things (IoT): A Vision, Architectural Elements, and Future Directions. *Future Generation Computer Systems* 29(7), 1645–1660.

Hassan, M., Hossain, E., and Niyato, D. (2013). Random Access for Machine-to-Machine Communication in LTE-Advanced Networks: Issues and Approaches. *IEEE Communications Magazine* 51(6), 86–93.

ISO. (2009). *ISO 31010:2009 Risk Management—Risk Assessment Techniques*. International Standard Organization.

Kandasamy, K., Srinvas, S., Achuthan, K., and Rangan, V.P. (2020). IoT Cyber Risk: A Holistic Analysis of Cyber Risk Assessment Frameworks, Risk Vectors, and Risk Ranking Process. *Journal on Information Security* 8, 1–18. https://doi.org/10.1186/s13635-020-00111-0.

Khan, R., Khan, S.U., Zaheer, R., and Khan, S. (2012). Future Internet: The Internet of Things Architecture, Possible Applications and Key Challenges. *10th International Conference on Frontiers of Information Technology*, Islamabad, Pakistan, 2012, pp. 257–260, https://doi.org/10.1109/FIT.2012.53.

Kumar, J.S., and Patel, D.R. (2014). A Survey on Internet of Things: Security and Privacy Issues. *International Journal of Computer Applications* 90(11), 20–26.

Kumar, N.M., and Mallick, P.K. (2018). The Internet of Things: Insights into the Building Blocks, Component Interactions, and Architecture Layers. *International Conference on Computational Intelligence and Data Science (ICCIDS 2018), Procedia Computer Science*, 132, 109–117.

Latifi, F., and Zarrabi, H. (2017). A COBIT5 Framework for IoT Risk Management. *International Journal of Computer Applications* 170, 40–43. www.ijcaonline.org/archives/volume170/number8/latifi-2017-ijca-914933.pdf.

Li, S., Xu, L., and Zhao, S. (2015). The Internet of Things: A Survey. *Information Systems Frontiers* 17(2), 243–259.

Madakam, S., Ramaswamy, R., and Tripathi, S. (2015). Internet of Things (IoT): A Literature Review. *Journal of Computer and Communications* 3, 164–173. http://dx.doi.org/10.4236/jcc.2015.35021.

Mashal, I., Alsaryrah, O., Chung, T.Y., Yang, C.Z., Kuo, W.H., and Agrawal, D.P. (2015). Choices for Interaction With Things on Internet and Underlying Issues. *Ad Hoc Networks* 28, 68–90.

Misal, D. (2019). Top 5 Skills Required to Become an IoT Developer. https://analyticsindiamag.com/top-5-skills-required-to-become-an-iot-developer/. Accessed 25 October 2022.

Nord, J.H., Koohang, A., and Paliszkiewicz, J. (2019). The Internet of Things: Review and Theoretical Framework. *Expert Systems with Applications* 133, 97–108.

Onibonoje, M.O., Nwulu, N.I., and Bokoro, P.N. (2019). An Internet-of-Things Design Approach to Real-Time Monitoring and Protection of a Residential Power System. *2019 IEEE 7th International Conference on Smart Energy Grid Engineering (SEGE)*, Oshawa, ON, Canada, pp. 113–119. https://ieeexplore.ieee.org/document/8859879.

Pallavi, S., and Smruti, R.S. (2017). Internet of Things: Architectures, Protocols, and Applications. *Journal of Electrical and Computer Engineering* 25. https://doi.org/10.1155/2017/9324035.

Ram, M. (2015). The Internet of Things: Solving Security Challenges from the Fringe to the Core. https://afilias.info/blogs/ram-mohan/internet-things-solving-security-challenges-fringe-core. Accessed 12 August 2022.

Rapid. (2021). 10 Essential Risk Management Skills That Every Manager Should Have. www.rapidglobal.com/knowledge-centre/10-essential-risk-management-skills-.that-every-manager-should-have/. Accessed 19 October 2022.

Ray, P.P. (2018). A Survey on Internet of Things Architectures. *Journal of King Saud University—Computer and Information Sciences* 30(3), 291–319.

Said, O., and Masud, M. (2013). Towards Internet of Things: Survey and Future Vision. *International Journal of Computer Networks* 5(1), 1–17.

Salamai, A.A. (2021). Risk Management Techniques on the Internet of Things. *Journal of Computer Science and Information Systems* 2(5).

Samaila, M.G., Neto, M., Fernandes, D.A.B., Freire, M.M., and Inacio, P.R.M. (2018). *Challenges of Securing Internet of Things Devices: A Survey*. Wiley, pp. 1–32. https://doi.org/10.1002/spy2.20.

Schiffman, J., Moyer, T., Jaeger, T., and McDaniel, P. (2011). Network-Based Root of Trust for Installation. *IEEE Security and Privacy* 9(1), 40–48.

Sengupta, J., Ruj, S., and Das Bit, S. (2020). A Comprehensive Survey on Attacks, Security Issues and Blockchain Solutions for IoT and IIoT. *Journal of Network and Computer Applications* 149, 102481.

Showail, A.J. (2021). *Internet of Things Security and Privacy*. ItechOpen. https://doi.org/10.5772/intechopen.96669.

Siboni, S., Sachidananda, V., Meidan, Y., Bohadana, M., Mathov, Y., Bhairav, S., Shabtai, A., and Elovici, Y. (2019). Security Testbed for Internet-of-Things Devices. *IEEE Transactions on Reliability* 68(1), 23–44.

Sundmaeker, H., Guillemin, P., Friess, P., and Woelffle, S. (2010). Vision and Challenges for Realizing the Internet of Things. *Cluster of European Research Projects on the Internet of Things, European Commission* 3(3), 34–36.

Theoleyre, F., and Pang, A.C. (eds). (2013). *Internet of Things and M2M Communications.* Gistrup, Denmark: River Publishers.

Tiganoaia, B., Niculescu, A., Negoita, O., and Popescu, M. (2019). A New Sustainable Model for Risk Management—RiMM. *Sustainability* 11, 1178. https://doi.org/10.3390/su11041178.

Vermesan, O., Friess, P., Guillemin, P., Gusmeroli, S., Sundmaeker, H., Basssi, A., and Doody, P. (2011). Internet of Things Strategies Research Road Map. *Internet of Things: Global Technological and Societal Trends* 1, 9–52.

Villamil, S., Hernandez, C., and Tarazona, G. (2020). An Overview of Internet of Things. *TELKOMNIKA Telecommunication, Computing, Electronics and Control* 18(5), 2320–2327. https://doi.org/10.12928/TELKOMNIKA.v18i5.15911.

Wu, M., Lu, T.J., Ling, F.Y., Sun, J., and Du, H.Y. (2010). Research on the Architecture of Internet of Things. *Proceedings of the 3rd International Conference on Advanced Computer Theory and Engineering (ICACTE' 10)*, IEEE 5, pp. 484–487.

Zanella, A., Bui, N., Castellani, A., Vangelista, L., and Zorzi, M. (2014). Internet of Things for Smart Cities. *IEEE IoT Journal* 1(1), 22–32.

Zhang, M., Sun, F., and Cheng, X. (2012). Architecture of Internet of Things and Its Key Technology Integration Based-on RFID. *5th IEEE International Symposium on Computational Intelligence and Design (ISCID)*, pp. 294–297. Accessed October 2012.

Zhao, K., and Ge, L. (2013). A Survey on the Internet of Things Security. *9th IEEE International Conference on Computational Intelligence and Security (CIS)*, pp. 663–667.

Zhou, A. (2016). *Analysis of the Volkswagen Scandal Possible Solutions for Recovery.* School of Global Policy and Strategy, Course on Corporate Social Responsibility. UC, San Diego. Retrieved from: https://gps.ucsd.edu/_files/faculty/gourevitch/gourevitch_cs_zhou.pdf

6 Blockchain Technology

6.1 INTRODUCTION TO AND COMPONENTS OF BLOCKCHAIN TECHNOLOGY

The term "blockchain technology" refers to a sophisticated data structure comprising a list of records called blocks (Bhavya et al. 2018). Ahmed et al. (2019) and Ayesha et al. (2019) opined that the blocks in a blockchain network consist of four essential components: data, a timestamp, a hash (an identifying number) of the previous block, and a hash of the current block, while every new block is linked to a previous or earlier block (Ahmed et al. 2019). Hence, most people associate blockchain technology with Bitcoin and other crypto currencies (Zhao et al. 2016). In addition to crypto currencies, this technology has been used in several other fields for about 10 years, especially in finance, higher education, healthcare, logistics, and commerce (David et al. 2022, 2023). Additionally, it might be considered a "Smart System" rather than a "Blockchain" in general (Alam and Benaida 2020; Mahmood et al. 2020). Blockchain Training Alliance (2017) claims that blockchain technology is a secure, sophisticated, and complex technology with advanced protocols for transacting and storing valuable items such as money, property, identification credentials, and contracts through the Internet without going through a third-party intermediary like a government or bank. Satoshi Nakamoto proposed blockchain technology from the Bitcoin cryptocurrency in 2008 (Ahmed et al. 2019). Mohamad et al. (2019) opine that blockchain technology can be used to protect patients' health information. Generally, there are three basic design types for blockchain technology: public (permissionless), private (permissioned), and hybrid blockchain (Bhavya et al. 2018).

Everybody has access to participate in public blockchain designs, which are open-source designs. This form of blockchain design uses the well-known and widely used proof-of-work framework or algorithm (Ahmed et al. 2019). Furthermore, every participant can see the details in a public blockchain transaction. No user or outside entity controls how the transaction is carried out (Bhavya et al. 2018). On the other hand, in a private blockchain design, users must request access before joining the network. Only authorized parties, in this case, have access to all transactions. However, a hybrid blockchain's design is flexible and allows for whether to make a given type of data public or private (Bhavya et al. 2018). Figure 6.1 depicts the structure of blockchain technology.

Furthermore, blockchain technology has fundamentally altered how the world evolves. For instance, a lot of paperwork and laborious work must be undertaken if a person wants to purchase an asset, such as a house, a car, or other things. But with blockchain, these time-consuming tasks are completed quickly and with ease (Miah et al. 2019). Bitcoin and other crypto currencies' infrastructure is based on blockchain technology. Wang and Jean-Phillippe (2017) define crypto currencies

DOI: 10.1201/9781003522102-8

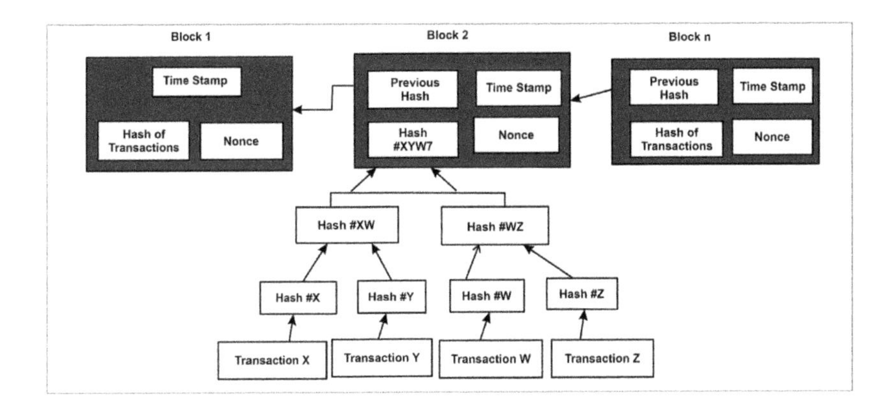

FIGURE 6.1 The structure of blockchain technology (Abu-elezz et al. 2020, p. 2)

as "digital tokens that may be transferred online, utilizing cryptographic hashes and digital signatures to validate transactions and prevent multiple spending of the same token." Crypto currencies, to put it very simply, are peer-to-peer networking technology and public key cryptography-based online digital payment systems. Transactions in this system are permanent, encrypted, and guaranteed to be recorded and maintained in a distributed ledger that guarantees the transaction's validity by all participants. Fortunately, blockchain has been recognized as a cutting-edge technology capable of revolutionizing the future of transaction-based exchanges, unlike how the World Wide Web and other modern media streaming services first emerged.

The critical distinction between early Internet protocols and blockchain is that while TCP/IP allowed for the instantaneous transmission of information, blockchain allows for the instantaneous transfer of money (Bheemaiah and Kamappa 2015). Moreover, there are three stages of blockchain technology. These stages correspond to blockchain versions 1.0, 2.0, and 3.0. The Blockchain 1.0 revolution is described as the online cryptocurrency phase developed and supported by the current Bitcoin system, whereby there are predictions by Blockchain developers that Blockchain 2.0 would soon have applications. But to do this, the blockchain must track and secure contracts, public records, property ownership, and financial information. However, databases with land ownership records resistant to fraud and errors are examples of Blockchain 2.0 systems (*The Economist* 2015).

Furthermore, it is predicted that Blockchain 3.0 will broadly impact education, science, and the medical field. However, it is anticipated that the distributed and open blockchain would automatically transfer private information maintained by institutions (Swan 2015). Therefore, White (2017) opines that blockchain technology is created as a decentralized network that users and intermediate organizations regularly verify. Blockchain, however, can generate benefits for businesses and the supply chain management and it is used within the energy value chain (Tapscott and Tapscott 2017; Damisa et al. 2022; Nwulu and Damisa 2023).

In their research, Damisa et al. (2022), Bocek et al. (2017), and Fairfield (2014) noted that blockchain could be used to build a smart contract, an independent contract made with the support of a third party or institution like banks, the government, or lawyers. Moreover, due to the need for middlemen or intermediaries to store these data, Nakasumi (2017) highlighted enterprise resource planning systems as being used for information exchange. Because it uses a decentralized network, blockchain technology can be used to reduce system dependencies and vulnerabilities (Bocek et al. 2017). However, Casey and Wong (2017) asserted that the blockchain might encounter supply chain delays. However, this may occur due to the tracing and tracking nature of blockchain (Glover and Hermans 2017), which can monitor the supply chain (Casey and Wong 2017). However, some shortcomings exist in blockchain adaptability into the supply chain due to the complex effort of gathering all partners to integrate their supply chain. In the study, Zheng et al. (2015) and Damisa et al. (2022) argued about the adaptability of blockchain regarding the method of handling online transactions to ensure robust security measures in the blockchain. Also, due to the lack of ability in some developing countries, Kshetri (2018) argued that blockchain experiences difficulties in the global supply chain. Figure 6.2 shows the application areas of blockchain technology.

But how does blockchain work? A blockchain is known as a chain of blocks consisting of three things: data, cryptographic hash, and hash of the previous block. In the chain, each block has a cryptographic hash of its down to the last block on the chain. Primarily, a block is the building unit of a blockchain that comprises well-defined data or information. This data or information is added to the block in the blockchain chronologically to create a chain of blocks linked together.

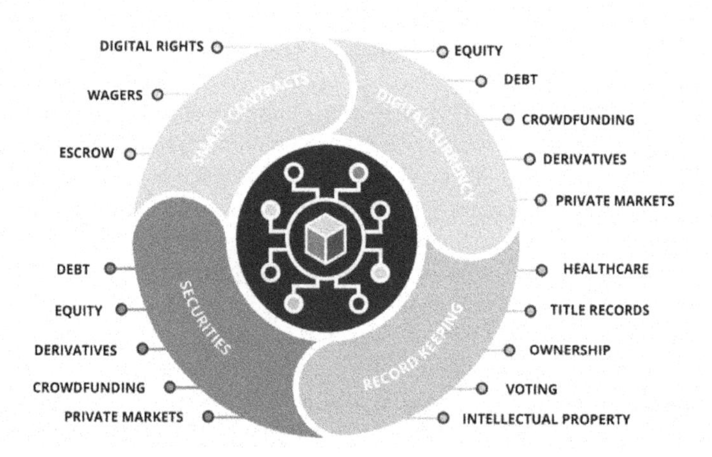

FIGURE 6.2 Application of blockchain in different areas (Patibandla and Vejendla 2022, p. 24)

Thus, a structured database of transactions that can be shared with different nodes such as computers or servers is formed in a secured network. The proof-of-work algorithm, which includes adding a unique number to a hashed or encrypted block, is also used to solve cryptographic problems for generating the next block in the chain. On the other hand, the distinctive alphanumeric identifying code or number known as a hash is produced whenever a transaction occurs in the blockchain. Based on its timestamp, prior blocks' hashes, and its contents, hash.

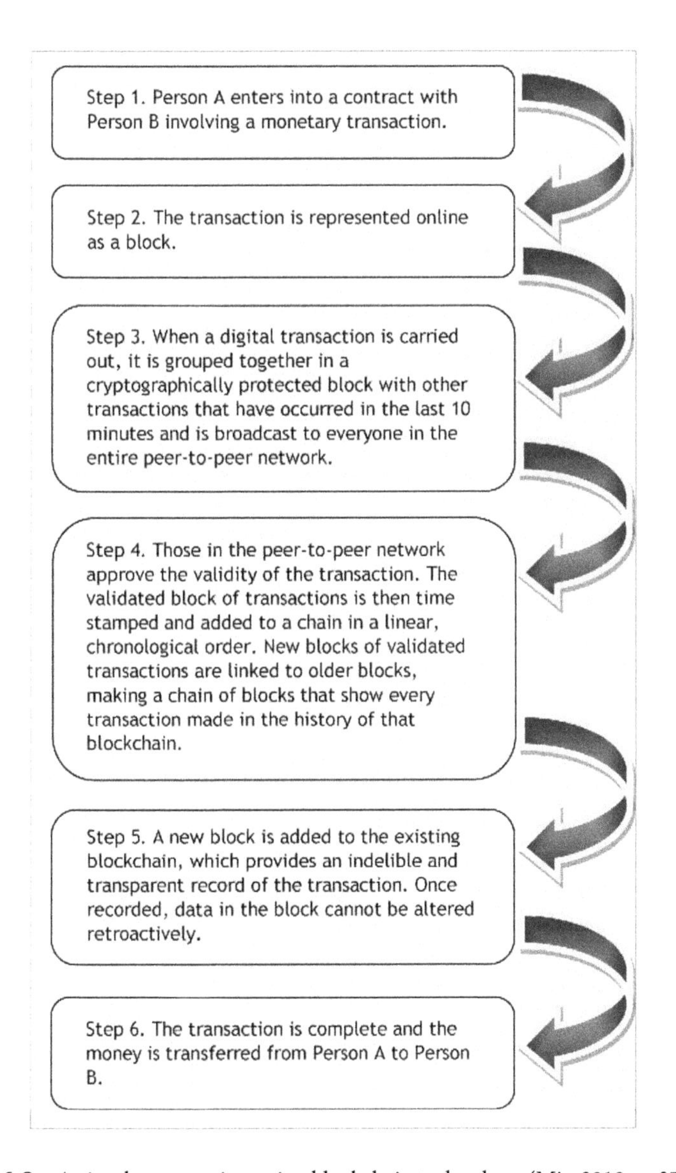

FIGURE 6.3 A simple transaction using blockchain technology (Min 2019, p. 37)

How blockchain technology functions is depicted in Figure 6.3. Every transaction that takes place on the blockchain is recorded in a block, which must then be confirmed and approved to be added to the chain. Before a block can be added to the chain of the distributed network, it must first be validated by many nodes (clients or servers) and the nodes with the highest stack in the chain. A distinct distinguishing code, or hash, is generated following the validation of each block. However, this does not require the intervention of any outside entity to validate or complete any transaction. The block number, height, or header hash can be used to identify individual blocks in the chain. The hash function is a computer technique to find block data. The hash function makes the blockchain's data unchangeable, ensuring participants cannot access it. Every block in a blockchain has its hash algorithm, and the recorded data is irreversible. A small change, however, results in the creation of a new block in the chain. Blockchain functions as a type of digital notary with time stamps to prevent data manipulation (Suman and Patel 2021).

According to Suman and Patel (2021), the characteristics of blockchain technology include the following:

1. A select group of legitimate individuals can write and read blocks on the blockchain, and all entries are permanent, transparent, and searchable.
2. The blockchain allows for the writing and reading of blocks, and all entries are permanent, visible, and searchable.
3. Every entry is permanent, transparent, and searchable in the blockchain, allowing only a small number of trusted individuals to write and read blocks.
4. In the blockchain, blocks can be created and read by specifically authorized individuals, and all entries are irreversible, transparent, and searchable.
5. It enables peer-to-peer value transfers without needing a central middleman like a bank or other financial institution.

6.2 RISK ASSOCIATED WITH BLOCKCHAIN TECHNOLOGY

The term "blockchain" was first used with the 2009 introduction of Bitcoin. It is frequently referred to as a distributed ledger and offers a secure cryptographic or signature mechanism-based open or decentralized system for exchanging data or digital assets. The development of smart contracts, a collection of open programs that can provide a secure environment and generate an event based on transactions made on the blockchain, was made possible by the advent of this technology. This decentralized software program can communicate with itself. However, the development of this new technology sparked increased attention among professionals and other technologies. But blockchain technology has advantages in terms of lowering attack risk and increasing effectiveness, particularly in situations involving trade reconciliation, clearing, and interbank payment systems. Blockchain technology has been used to enable the concurrent operation of the financial markets.

On the other hand, it is necessary to identify the risks connected to blockchain technology to ensure the business runs smoothly (Fennel 2019).

However, according to Fennel (2019), a 51% vulnerability attack is one of the dangers connected to blockchain technology. This gives an opportunity for hackers to assault the system, disable its functioning, and take over the blockchain. Launching a 51 percent assault makes it possible to manipulate and change data on the blockchain, start doubling every transaction, reverse those transactions, and obstruct mining activities and confirmation procedures. Blockchain has also been shown to be useful, although many hazards are involved. These risks can be connected to blockchain directly or indirectly through implementation, investment, legal, operational, security, finance, and other important aspects. Lack of standardization, difficulties integrating protocols, and subpar cryptocurrency appraisal are some of the general blockchain dangers.

On the other hand, the development risks of blockchain include a lack of standards, laws governing data protection, and a high energy demand. Furthermore, other blockchain security concerns include human-related risks, risks resulting from using public and private keys, vendor risks, and risks associated with untested programs. In addition, the legal hazards connected to blockchains include those related to data protection, jurisdiction, and dispute resolution (Iredale 2021).

Furthermore, there are ten related blockchain risk factors. These factors include the following:

- Key management
- Data management
- Performance and scalability
- Use ease of applicability
- Chain protection
- Integration and interoperability
- Regulations and compliance
- Disaster recovery
- Privacy and chain management.
- Network and consensus management

(Iredale 2021)

Consequently, the blockchain hazards area must be appropriately considered when developing a blockchain-related application (Iredale 2021). However, the risk associated with blockchain is discussed in different categories.

6.2.1 The General Risks Associated With Blockchain Technology

Lack of standardization: Despite blockchain technology's exciting potential, not everyone has adopted or utilized it. Lack of uniformity in standards is one of the causes (Institute 2019; Deshpande et al. 2017). If implemented, standardization will, in the opinion of experts, enhance the process of implementing technology, lower transaction costs, reduce regulatory risks, improve the interoperability of systems, and foster effective communication among market participants. It will also increase the likelihood of securing assets on the blockchain (Frank Cerveny 2019). However, there are no guidelines for using blockchain technology (König et al. 2020). As a result, the absence of standards is to blame for the barriers to adopting blockchain technology, as standardization is crucial in conventional IT (Deshpande et al. 2017). Worldwide communication relies heavily on standards; therefore, it would not be possible without established practices or procedures (Hurd and Isaak 2008). The need for proper protective measures to be adopted increases as sensitive and complicated information on IT systems does, which is accomplished through standardization (Meyers and Rogers 2004; Disterer 2013; Boehemer 2009).

Difficulties in integrating protocols: Protocols are essential elements of blockchain technology that enable secure information to be automatically shared between cryptocurrency networks. The blockchain's data is organized through protocols, and a defense mechanism is set up to stop attackers from causing harm to the blockchain. However, integrating blockchain protocols into a project is challenging due to the different API (application programming integration); implementing various blockchain initiatives is laborious. For instance, an integration layer for controlling these two enterprise systems is required to transmit information from the Hyperledger Fabric protocol to the Ethereum Protocol (Iredale 2021).

Poor evaluation of cryptocurrency: One of the biggest dangers of blockchain technology is the ephemeral nature of crypto currencies. Cryptocurrency price fluctuations with subsequent losses cause the markets to perceive blockchain differently. Additionally, the cryptocurrency's price dribble can potentially bankrupt thousands of investors. As a result, anyone who uses or depends on blockchain-based initiatives or crypto currencies risks losing money. This includes participants and third parties like banks (Iredale 2021).

6.2.2 Blockchain Development Risks

The risks associated with blockchain development include the following:

Underdeveloped standards: Standards are crucial to any emerging technology's success. Interoperability, trust building, and effective use of technology are all ensured by the proper standards being set at the appropriate stage of technological development. Standards facilitate technology creation and widespread use (European Commission 2022). However, due to its rapid expansion, blockchain has fallen short of establishing appropriate standards. As a result of the work being done by various organizations on their own distributed ledger technologies or blockchains, it is impossible to have widely used standards. The ferocious competition rate among diverse businesses hampers blockchain technology's major objective and standards creation. As a result, there are security, interoperability, and privacy problems (Iredale 2021).

High energy demand: One of the cutting-edge technologies that have been used in business, education, industries, and so forth is blockchain technology. The invention of bitcoins sparked the establishment of blockchain, a system for securely storing and exchanging digital information without the intervention of a third party. It has been utilized for purposes beyond being only a distributed ledger technology. However, there are some problems with blockchain technology. High energy consumption is one such concern. Because of the method employed to create them, it has been discovered that blockchains use a lot of energy (Ghosh and Das 2020).

Data privacy legislation: Data privacy is one of the problems with blockchain technology or distributed ledger technology. Blockchain data privacy standards must be enforced, similar to how numerous nations and regions have applied the General Data Protection Regulation of the European Union (Iredale 2021).

6.2.3 Blockchain Security Risks

The distributed ledger technology known as blockchain has built-in security mechanisms that make it resistant to hacking. But even with these characteristics, there are security dangers, therefore, it is not immune (Zamani et al. 2018). As is common with innovative technologies, cyber security risks and weaknesses have affected blockchain (Abdelwahed et al. 2020). According to the generations of blockchain technology, its cyber security risks are divided into three categories (Hasonova et al. 2019).

General security risks of Blockchains 1.0 and 2.0

- Double spending
- The 51% attack or Goldfinger attack
- Wallet security (private key security)
- The specific flaw in PoS
- Network-level attack
- Malleability attack
- Real DoS attack against the Ethereum network
- The specific flaw in DPoS
- Block produces collusion
- Exploit low voter turnout
- Attacks at scale

Blockchain 2.0 vulnerabilities

- Re-entrance vulnerability (DAO attack)
- Party multi-sig wallet
- King of the ether throne
- Governmental

Blockchain 3.0 vulnerabilities

- The attack against hyper ledger fabric
- General risk on private blockchain implementation

The above vulnerabilities lead to the execution of various security threats to the functionality of blockchain platforms. Figure 6.4 shows the blockchain vulnerabilities.

6.2.4 Legal and Compliance Risks of Blockchain Technology

The widespread adoption of blockchain technology, considering its risks to the financial system, especially cryptocurrency, has highlighted the risks associated

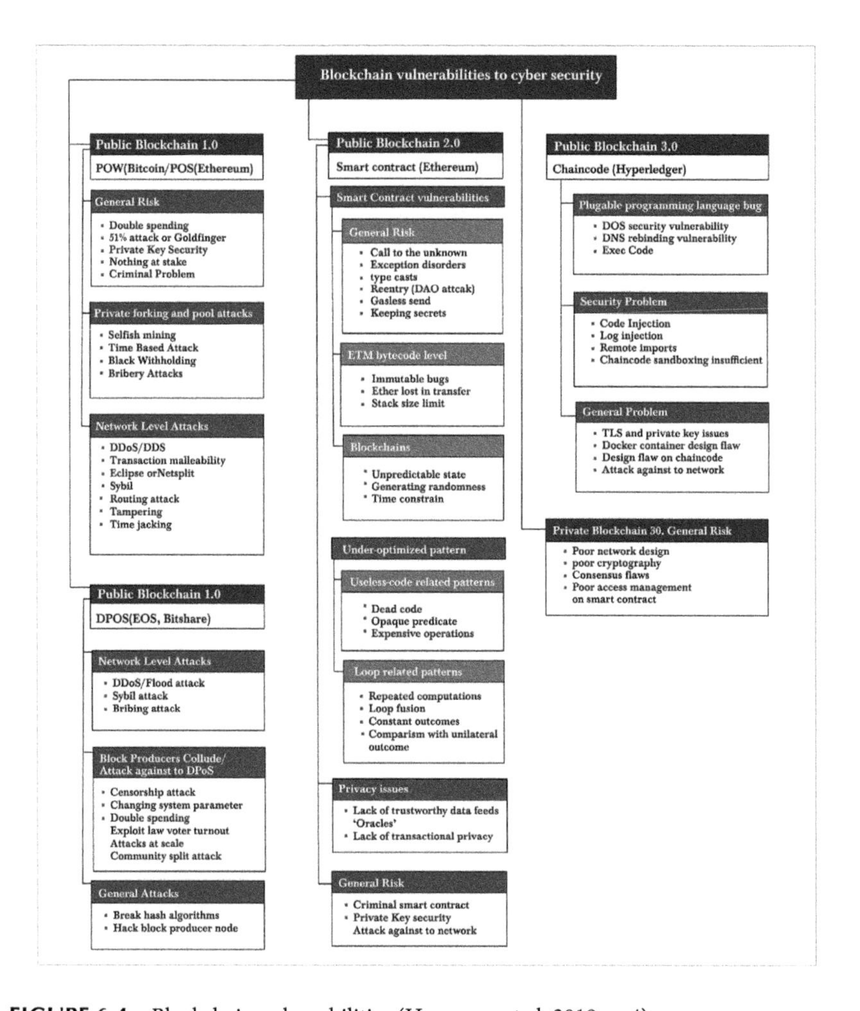

FIGURE 6.4 Blockchain vulnerabilities (Hasonova et al. 2019, p. 4)

with compliance and the necessary legal issues surrounding it, considering issues of the legal classification of digital assets, smart contracts, and security law issues among others. Hence, the General Data Protection Regulation (GDPR) in the United Kingdom has always been used as a measure of control. Meanwhile, any company is affected by the GDPR infringement, not just businesses in the EU. The distinction between a public and private blockchain is considered for compliance concerns with the EU's GDPR. The private controller of the data in the private blockchain must adhere to the necessary data protection requirements, such as authorizing access to the data, allowing for data rectification, and allowing data transfer. Additionally, a permissioned distributed ledger has higher compliance risks than a permissionless one. The dangers associated with GDPR compliance arise the moment third-party data is kept on the ledger (Cloots 2018).

6.2.5 Operational and Systemic Risks of Crypto Currencies and Blockchain

The promise of a decentralized payment system independent of third parties led to the popularity of crypto currencies. Because it is decentralized, a wide network of networks shares each transaction. However, as shown in Figure 6.5, mining has become highly concentrated (Lai et al. 2021, p. 34).

Voting power is concentrated when mining power is concentrated. Meanwhile, the blockchain's governance is determined by 50% + 1. This shows that a small group of the biggest miners may change the blockchain's algorithm to their favor. However, the concentration of mining power undermines the blockchain's internal

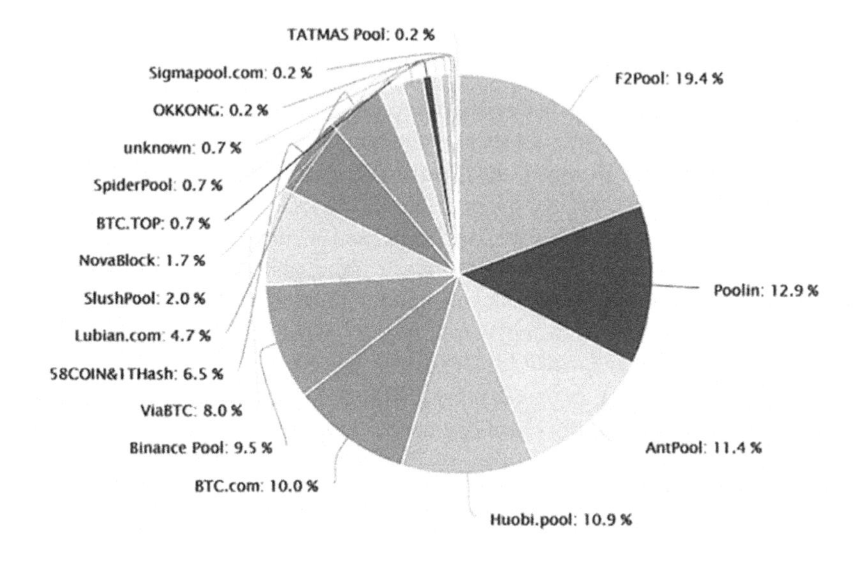

FIGURE 6.5 Concentration of miners (Lai et al. 2021, p. 34)

workings. The concentration of mining power increases the chance that a gang of miners will alter the algorithm and harm the company (Cloots 2018).

Moreover, blockchain technologies expose institutions to risks in tandem with current business processes. These risks include the following:

Business continuity risk: The distributed nature of blockchain technology results in redundancy, which makes it usually resilient. Meanwhile, hackers and technical hiccups can affect company processes built on blockchain technology. However, businesses must have a long-term business continuity plan and governance framework to reduce these risks (Deloitte 2017).

Reputational risk: The fundamental technological infrastructure must continue to function with legacy infrastructure that has existed for a long time. This includes blockchain technology. Reputational problems and a poor client experience emerge from the inability to do so (Deloitte 2017), thus impeding the reputation of the business and organization.

Information security risk: Blockchain technology completely supports transaction security but does not provide account or wallet security. Additionally, there are more significant cyber security threats in the blockchain network if an attacker takes advantage of around 51% of the network for a specific period, often in a public blockchain framework (Deloitte 2017).

Operational and IT risks: Existing policies and procedures must reflect new business processes. This is as speed, scalability, and the technology's interface with legacy systems could be additional implementation-related challenges (Deloitte 2017).

6.3 RISK MANAGEMENT TECHNIQUES FOR BLOCKCHAIN TECHNOLOGY

Blockchain is a decentralized technology that comprises a peer-to-peer network of information that stores digital assets and performs transactions based on distributed ledger technology. However, the records in the blockchain are decentralized; that is, they cannot be controlled by an external entity such as banks, government, and so forth. Hence it can mitigate risks associated with the third-party intervention, including hacking, compromised privacy, vulnerability to political turmoil, costly compliance with government rules and regulation, instability of financial institutions, and contractual disputes. Additionally, blockchain technology aids in the reduction of covert and invisible hazards that are difficult for a select few actors (such as the vendor, buyer, or financial institution) to identify during ordinary commercial transactions or supply chain operations. In other words, blockchain technology enables users to take advantage of various layers of protection. Figure 6.6 contrasts traditional risk management concepts with new risk management principles offered by blockchain technology (Min 2019).

Additionally, blockchain technology is applied to asset monitoring as a publicly accessible ledger that tracks and permanently records each supply chain

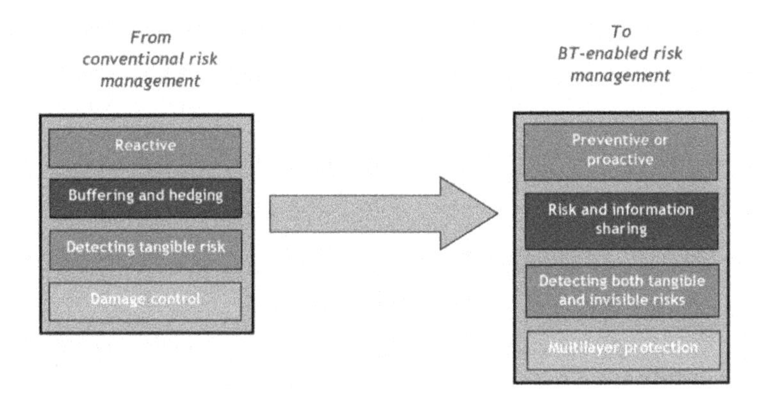

FIGURE 6.6 Evolution from conventional risk management to blockchain technology–enabled risk management (Min 2019, p. 42)

transaction for a specific commodity. The data in each blockchain block cannot be changed, making it impossible to create ownership history. Blockchain thus makes it simpler to track digital assets or items as they move from one location to another in the supply chain and prevents the transaction of fraudulent or counterfeit assets (Higgins 2017). Due to blockchain technology's reliance on cryptographic signatures, which makes it difficult for anyone to alter the data on the chain, this entire process led to it (Green 2017; Hackett 2017b). Blockchain networks can be encrypted to increase the security of blockchain even more (Min 2019).

Moreover, there are different categories of a risks management framework for managing the risks of blockchain, which are succinctly described as follows:

Blockchain general risk management process: A broader perspective of the blockchain technology cyber security risk management plan is as follows:

- Risk analysis
- Risk assessment
- Risk mitigation
- Risk monitoring (Kabanda 2021)

The Risk management process is illustrated in Figure 6.7.

Risk analysis/identification: Risk is the likelihood that a catastrophic occurrence may prevent one from achieving one's goals. In a system, the risk is determined or assessed in terms of the severity and likelihood of an event (Misra 2016). It is a foundation for making wise decisions in uncertain circumstances. Because of the increased concern about industrial and technical risks, it becomes essential in policy-making. Four methodologies are taken into consideration to

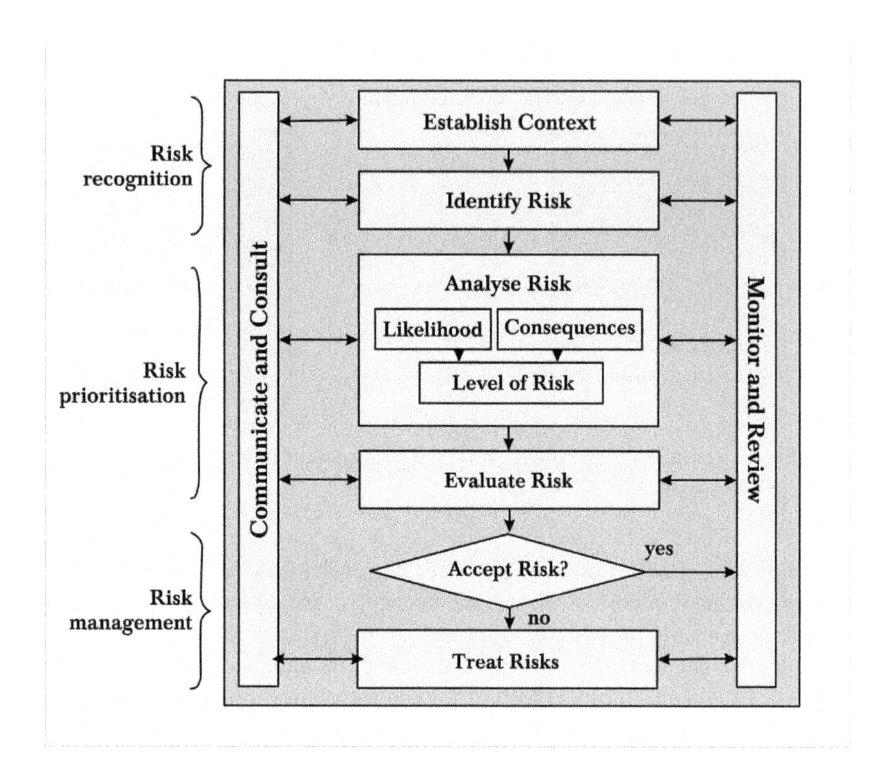

FIGURE 6.7 Risk management process (Kabanda 2021, p. 11)

grasp risk analysis properly. These consist of risk estimation with an engineering perspective, risk evaluation with decision analysis and risk perception, and risk management with policy analysis. Each of these methods has a unique viewpoint on the risk analysis processes. Additionally, there are five steps to complete risk analysis tasks. They consist of the following:

i. defining which outcomes could be labeled "adverse" or "beneficial"
ii. choosing which factors are to be given priority in the analysis
iii. assessing the magnitude of harm to which the public may be exposed
iv. Calculating the probabilities of various outcomes
v. Determining who will be affected by the risk

Risk assessment: The emergence of blockchain technology has completely changed the nature of business. This is due to its ability to undertake transactions and create records without the involvement of outside parties. Indeed, it has benefited companies and academic institutions and offers a wealth of opportunities. It may also expose institutions and other participants to new risks that must be identified, understood, eliminated, or reduced (Morganti et al. 2018). These important risks include technology, interoperability, data security, and third-party vendors. Consequently, it is also necessary to understand how to identify the dangers associated with blockchain technology (White et al. 2020). Therefore, they have to be assessed individually and collectively, determining the attack level, severity, and impact on a particular system. This may be done within a technical unit, through interdepartmental assessment, or through an outside consultant.

Risk mitigation: Numerous businesses and industries have embraced blockchain technology for three reasons: to improve value production, fortify their current value ecosystem, or establish new value ecosystems. Although blockchain technology has many advantages, it is also subject to various strategic risks, such as commercial, technological, and legal concerns. Therefore, before using blockchain technology to create a unique value, it is important to understand the risks that could arise and how to reduce them. However, the strategic mitigation approach includes the walled garden strategy, the many gardens approach, the choices approach, and the all-industry approach (Malhota et al. 2022).

Risk monitoring: This entails monitoring identified risks and the mitigation strategies deployed to manage blockchain risks. This can be done via intelligent agents, systems, and expert setup, especially for a huge volume of dynamic information in a distributed system (Wang et al. 2002). Blockchain developers should avoid the temptation of complacency after deploying a risk mitigation strategy but should continuously monitor the performance of such a strategy to align possible deviations with intention.

Furthermore, the following are the components of risk management approaches for cyber security in blockchain technology:

Affected systems are identified, protected, detected, responded to, and recovered using the NIST Cybersecurity Framework.

Risk categories: This focuses on five different categories of risk, including cyber risk, reputational risk, information risk, business resilience, and regulatory risk.

Governance risk and compliance

Third-party risk management

Policies, standards, and procedures

Risk evaluation (identifying, assessing, developing controls and making decisions, implementing controls, supervising, and evaluating) (Kabanda 2021).

Multi-criteria decision analysis/making: Over the past seven decades, multi-criteria decision analysis has been applied to different technological solutions. It has also been used in several other application fields. It involves the analysis of different choices in making informed decisions, especially in analyzing different solutions against blockchain risks. Risk-based decisions in information security involve risk metrics linked to the triplet of threat, vulnerability, and consequences (Alexander et al. 2017). Thus, this method can classify risk as either simple or complicated and unclear, thereby enhancing the decision regarding the mitigation and management strategy to adopt and enforce.

Risk smart ledgers: During the process of identifying risks, a risk register table is created and reserved in the smart risk contracts. Risk control and analysis are used to update the contracts. However, the three main categories of risky smart contracts are as follows. The first is the summarized risk item ledger, which contains all of the information on hazards from phase of identification through the phase of therapy. The risk assessment ledger follows, which includes comprehensive data on the risk assessment of various risk items. The risk response ledger, which stores the account for various response programs under one of the identification pieces, is the third. The three types of ledgers are depicted in Figure 6.8 (Ma et al. 2018).

Within the ledgers, some attributes make up the ledger. These attributes are discussed in the following:

Risk event ID: Risk managers use this to identify each risk event. It can be in the form of unique numbers.

Name: This represents the name of each detected risk, such as server failure, a good reputation, and so forth.

Description: This attribute provides a detail description of a risk event.

Category: It classifies the risks into different categories. For instance, server failure can be categorized into hardware technology.

Trigger: It is a sign of the occurrence of a risk event.

Status: The status of the risk is the current formation of a risk.

Grade: This is often seen as a number that indicates the level of risk events.

Probability: This is known as the potentiality of risks to occur.

Influence: This attribute shows that the occurrence of risks event will impact achieving the aim of a system.

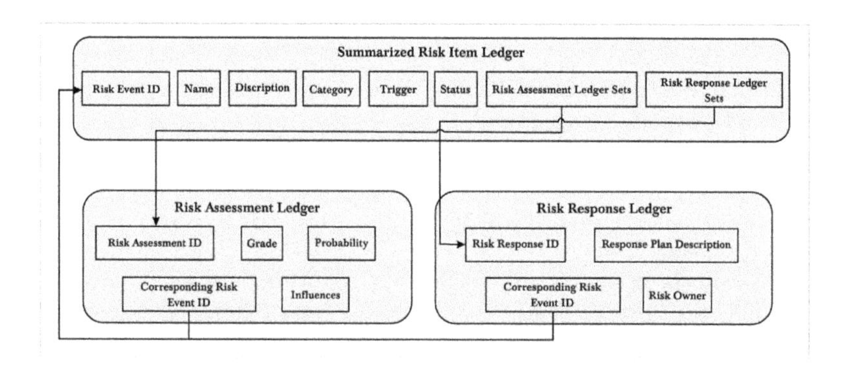

FIGURE 6.8 Three types of smart ledgers (Ma et al. 2018, p. 3)

Response plan description: a detailed response plan is formulated.
Risk owner: This represents an individual responsible for any related risk events and creates coping mechanisms.

Risk association tree: The Merkle tree or a hash tree must be used to determine the relationship between the three types of ledgers in the risk register table explained earlier. This risk association tree makes it easier to rapidly find the most recent data on the risk register. The three risk Merkle trees, on the other hand, were created in accordance with the properties of blockchain, which include the following:

Transaction tree: This feature holds each transaction in a unique block.
Receipt tree: It consists of several pieces of data showing the overall impact of each transaction.
Risk association tree: This feature holds the relationship between the three types of ledgers.

The first two of the three trees mentioned earlier are now used in blockchain to minimize storage costs and guarantee data consistency. The three types of ledgers are connected using the hash function by the risk association tree. The risk association tree is displayed in Figure 6.9.

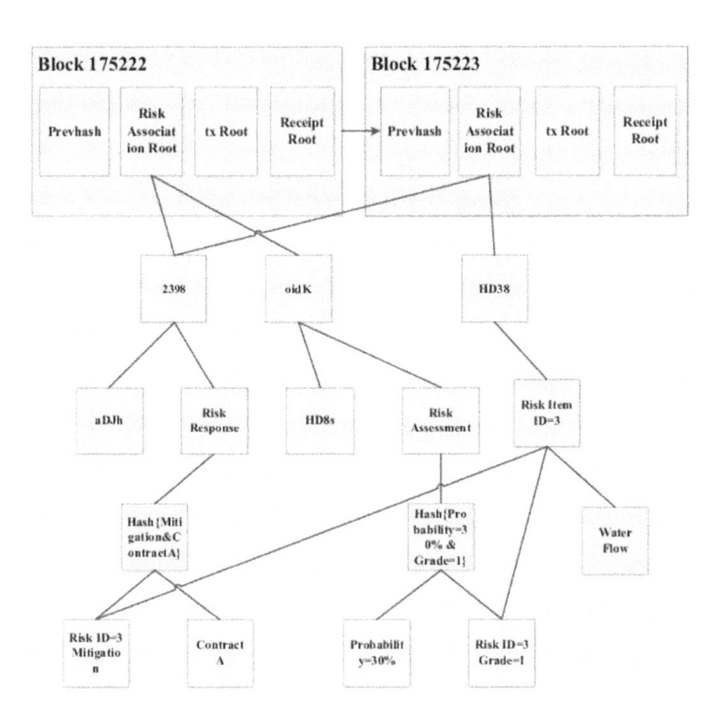

FIGURE 6.9 Risk association tree (Ma et al. 2018, p. 4)

The risk identification ledger is in Block 175222 in Figure 6.9, whereas Block 175223 houses the risk response ledger and the risk item ledger. The hash values, however, use the risk item, risk assessment, and risk response. Thus, the tree node relationship establishes an association relationship with the "grade" and "mitigation" attributes. Each blockchain agent can now download the block header because of this relationship being created. However, the Merkle tree can identify the risk response and risk assessment data in Block 175222 with the help of the hash value of the risk assessment ID and risk response ID in the "Risk Item ID=3" ledger on Block 175223. Similarly, the hash value can be used to find the information in the ledger marked "Risk Item ID=3" on Block 175223 for the known corresponding risk item ID in the risk assessment ledger in Block 175222 (Ma et al. 2018).

Risk management of crypto-assets: In the modern world, where technology has largely taken over the financial industry and all other sectors, it is essential to assess the framework and existing risk management approaches before incorporating them into the risk management of cryptographic assets. However, managing crypto assets is a component of the economic security system's risk management culture. Figure 6.10 illustrates crypto assets' proposed risk management culture (Bashynska et al. 2019).

Ontology-based reference model for security risk management of the blockchain-based applications: The reference model consists of three main components: the definition of settings, risk assessment and analysis, guidelines and controls for security risks, and countermeasures.

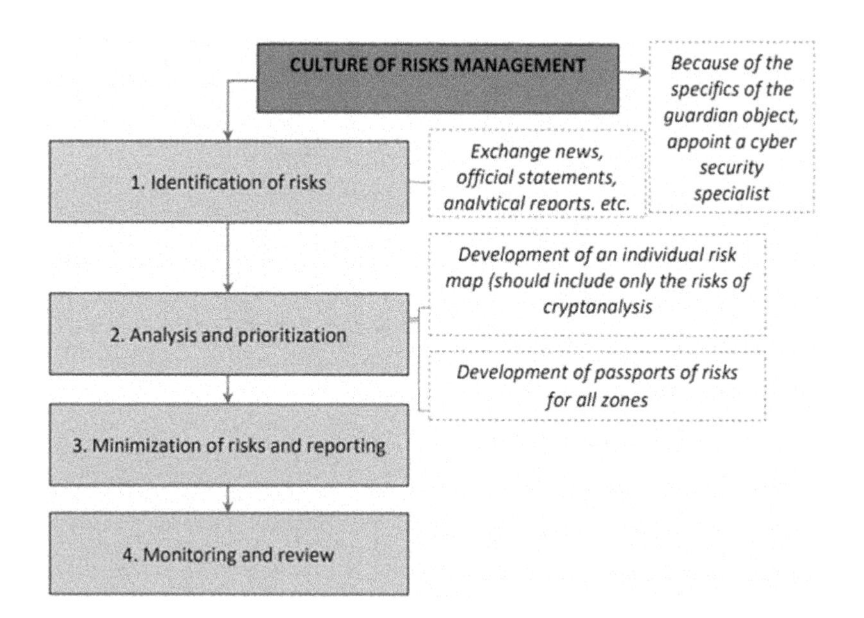

FIGURE 6.10 Risks management culture (Bashynska et al. 2019, p. 1130)

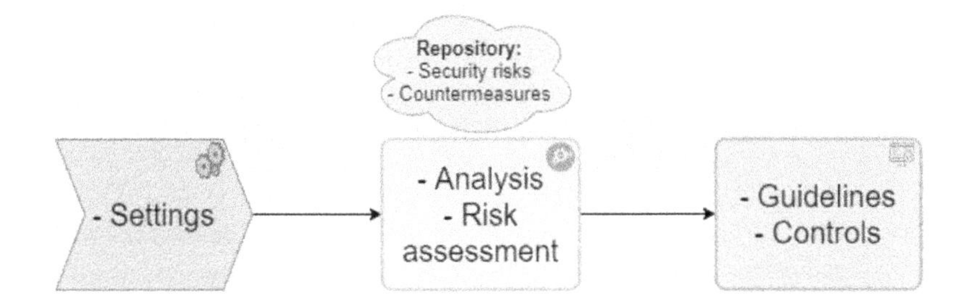

FIGURE 6.11 Reference model for security risk management of blockchain-based application (Iqbal 2020, p. 6)

Furthermore, the analysis and risk assessment components contain "a repository of security risks and countermeasures." Figure 6.11 presents the abstract representation of the reference model.

6.4 RISK MANAGEMENT SKILLS FOR BLOCKCHAIN TECHNOLOGY

Tucci (2022) defines risk management as recognizing, evaluating, and controlling threats to an organization's assets and revenue. Several things, such as monetary unpredictability, legal responsibilities, technological problems, strategic management blunders, accidents, and natural calamities, can cause these dangers. Furthermore, as defined by Misra (2016), risk management is a process that includes risk recognition, risk assessment, the creation of management strategies, and the use of managerial resources to reduce risks. However, one of the goals of risk management is to lower various hazards to a level that society will tolerate, which will require a skill set for blockchain technology, as skills are performance enhancement for risk management. Since the inception of the bitcoin cryptocurrency, risk and risk management have been of great concern to the public. Still, there has been an exponential increase in companies introducing blockchain technology into their business model. During this process, it is possible to introduce a new level of risk due to the absence of generally accepted standards guiding its development (Willis Towers Watson 2019). This will entail optimizing the skills needed for blockchain technology development and managing the risks that might arise from blockchain utilization and development. However, the risk management skills in blockchain technology can be either soft or technical skills. Following are the primary blockchain developer skills that one must have to be able to manage blockchain development.

Cryptography: The emergence of disruptive technologies brought about the widespread of cryptography. However, with the advent of electronic communications on computer networks, there is a need to ensure that conversations and transactions are done confidentially. Hence, cryptography is extensively applied. This technology uses encryption techniques (Onwutalobi 2011).

In-demand software development skills: Blockchain development is vast and therefore requires the knowledge of the following languages.

i. **Java:** This language is well-known among blockchain developers. It is simple, easy to learn, and capable of developing complex solutions. A popular cryptocurrency known as NEM is based on Java (Sahu 2022)
ii. **C++:** It is a popular language among blockchain developers who can perform many blockchain-related tasks. The creators created the Bitcoin core with C++. Hence, it is a technical skill for blockchain developers (Sahu 2022)
iii. **Python:** Python is a programming language that can be used for both front-end and back-end development. It has a lot of libraries. It has a huge community of about 4 million developers that can help with complex problems. Also you can learn Ruby, C#, and JavaScript. Most importantly, one's ability to learn and work in multiple languages is one of the vital blockchain developer skills (Sahu 2022).

Smart contracts: A self-executing contract known as a "smart contract" uses blockchain technology to digitally enforce, verify, or facilitate contract performance or negotiation. Because blockchain is decentralized, smart contracts guarantee the legitimacy of transactions between any contracting parties (Nzuva 2019).

Data structures: A crucial foundational skill in using information in computer science and other fields is data structure. It is a heavy burden for developing an individual's computational thinking and computer program design abilities. There are three layers to this technical skill: conceptual representation, building data models, and designing algorithms (Yu et al. 2019). Because the entire blockchain network is built up of data structures, it is crucial to possess this ability (Sahu 2022).

Blockchain architecture: To become a blockchain developer, one must be familiar with blockchain architecture. That is, you must know the ledger, consensus, and how smart contract works. However, blockchain architecture is categorized into three. These include consortium, private, and public architecture (Sahu 2022).

Web development: Becoming a blockchain developer means you might usually be developing web applications. Hence, learning how to design websites will enable you to fully understand how to create complex web applications that use blockchain technology (Sahu 2022).

Consequently, the soft skills required to learn for the management of risks in blockchain technology include the following:

Problem-solving skill: The ability to identify problems, brainstorm, and analyze answers and implement the best solutions. As of 2022, one of the sought-after skills is problem-solving skills (Kaplan 2022).

Critical-thinking skill: The capacity to reason logically and systematically to grasp how concepts or facts are related to one another is known as critical thinking. In other words, it involves "thinking about thinking"—finding, examining, and resolving problems with our thought processes (Tomaszewski 2022).

Communication skills: To speak effectively and maintain eye contact with a wide range of people, use a diverse vocabulary, adapt your language to your audience, listen attentively, and effectively present your ideas, you need communication skills.

Teamwork: This is the ability to collaborate with a group to complete a task. Due to the complex nature of blockchain technology, one must be willing to work effectively and efficiently with a team.

Adaptability: As technology advances, firms adapt to new trends and changes. Hence, the ability to quickly adapt to changes within the blockchain domain is essential.

Organizational skill: Organizational skill combines problem-solving, critical thinking, and adaptability skills. Organizational skills require the fusion of several elements to reach the desired goal.

6.5 SUMMARY

This chapter examined the various risks of blockchain technology and its different categories. However, different risk management techniques in blockchain technology were discussed in the chapter. First and foremost, these risks can be managed or reduced through the risk management process, including risk analysis, risk assessment, risk monitoring, and risk mitigation. The next techniques that were outlined are the multi-criteria decision analysis/making, risk smart ledgers, risk association tree using the Merkle tree, risk management of crypto-assets, and the proposed ontology-based reference model for security risk management of blockchain-based applications. Also, the risk management skills for blockchain technology that must be learned were categorized into soft and technical skills. The technical skills include cryptography, in-demand software development skills (Java, C++, Python, etc.), Smart contracts, data structures, blockchain architecture, and web development. At the same time, soft skills include problem-solving, critical thinking, communication, teamwork, adaptability, and organizational skills.

REFERENCES

Abdelwahed, I.M., Ramadan, N., and Hefny, H.A. (2020). Cybersecurity Risks of Blockchain Technology. *International Journal of Computer Applications* 177(42), 975–8887.

Abu-Elezz, I., Hassan, A., Nazeemudeen, A., Househ, M., and Abd-Alrazaq, A. (2020). The Benefits and Threats of Blockchain Technology in Healthcare: A Scoping Review. *International Journal of Medical Informatics* 142, 104246. https://doi.org/10.1016/j.ijmedinf.2020.104246.

Ahmed, A.M., Olov, S., and Karl, A. (2019). A Survey of Blockchain from the Perspective of Applications, Challenges, and Opportunities. *IEEE Access* 7, 117134–117151.

Alam, T., and Benaida, M. (2020). Blockchain and Internet of Things in Higher Education. *Universal Journal of Educational Research* 8.5, 2164–2174. https://ssrn.com/abstract=3638997 or http://dx.doi.org/10.2139/ssrn.3638997.

Alexander, A.G., Phuoc, Q., Mahesh, P., Zachary, A.C., Jeffrey, M.K., Dayton, M., and Igor, L. (2017). Multicriteria Decision Framework for Cyber Security Risk Assessment and Management. *Wiley Online Library*. https://doi.org/10.1111/risa.12891.

Ayesha, S., Usman, Q., and Ayesha, K. (2019). Using Blockchain for Electronic Health Records. *IEEE Access* 7, 14.

Bashynska, I., Malanchuk, M., Zhuravel, O., and Olinichenko, K. (2019). Smart Solutions: Risk Management of Crypto-Assets and Blockchain Technology. *International Journal of Civil Engineering and Technology (IJCIET)* 10(2), 1121–1131. https://iaeme.com/MasterAdmin/Journal_uploads/IJCIET/VOLUME_10_ISSUE_2/IJCIET_10_02_108.pdf.

Bheemaiah, K. (2015). Block Chain 2.0: The Renaissance of Money. *WIRED*. www.wired.com/insights/2015/01/block-chain-2-0/. Accessed 6 January 2015.

Blockchain Training Alliance. (2017). Blockchain: What is Blockchain? www.blockchaintrainingalliance.com. Accessed 23 July 2022.

Bocek, T., Rodrigues, B.B., Strasser, T., and Stiller, B. (2017). Blockchains Everywhere: A Use-Case of Blockchains in the Pharma Supply Chain. *2017 IFP/IEEE Symposium on Integrated Network and Service Management (IM)*, pp. 772–777.

Boehemer, W. (2009). Appraisal of the Effectiveness and Efficiency of an Information Security Management System Based on ISO 27001. *Proceedings of the 2008 Second International Conference on Emerging Security Information, Systems and Technologies*, Darmstadt, pp. 224–231.

Casey, M.J., and Wong, P. (2017). Global Supply Chains Are About to Get Better, Thanks to Blockchain. *Harvard Business Reviews* 13, 1–16.

Cloots, A.S. (2018). *Cryptocurrencies, Blockchain and Risk Management: Legal, Operational and Systemic Risks*. Cambridge: University of Cambridge, Judge Business School, pp. 1–10.

Damisa, U., Nwulu, N.I., and Siano, P. (2022). Towards Blockchain-Based Energy Trading: A Smart Contract Implementation of Energy Double Auction and Spinning Reserve Trading. *Energies* 15(11), 4084. https://doi.org/10.3390/en15114084.

Damisa, U., Oluseyi, P.O., and Nwulu, N.I. (2022). Blockchain-Based Gas Auctioning Coupled With a Novel Economic Dispatch Formulation for Gas-Deficient Thermal Plants. *Energies* 15(14), 5155. https://doi.org/10.3390/en15145155.

David, L.O., Nwulu, N.I., Aigbavboa, C.O., and Adepoju, O.O. (2022). Integrating Fourth Industrial Revolution (4IR) Technologies into the Water, Energy & Food Nexus for Sustainable Security: A Bibliometric Analysis. *Journal of Cleaner Production* 363, 132522. https://doi.org/10.1016/j.jclepro.2022.132522.

David, L.O., Nwulu, N.I., Aigbavboa, C.O., and Adepoju, O.O. (2023). Resource Sustainability in the Water, Energy, and Food Nexus: Role of Technological Innovation. *Journal of Engineering, Design, and Technology*. https://doi.org/10.1108/JEDT-05-2023-0200.

Deloitte. (2017). Bockchain Risk Management: Risk Functions Need to Play an Active Role in Shaping Blockchain Strategy. Deloitte, London England. Retrieved from: https://www2.deloitte.com/content/dam/Deloitte/us/Documents/financial-services/us-fsi-blockchain-risk-management.pdf

Deshpande, A., Stewart, K., Lepetit, L., and Gunashekar, S. (2017a). *Distributed Ledger Technologies/Blockchain: Challenges Opportunities and the Prospects for Standards; Overview Report*. London: The British Standards Institution (BSI).

Deshpande, A., Stewart, K., Lepetit, L., and Gunashekar, S. (2017b). *Understanding the Landscape of Distributed Ledger Technologies/Blockchains: Challenges, Opportunities, and the Prospects for Standards*; Technical Report. London: British Standards Institutions.

Disterer, G. (2013). ISO/IEC27000, 27001, and 27003 for Information Security Management. *Journal of Information Security* 4, 92–100.

The Economist. (2015). The Great Chain of Being Sure About Things. www.economist.com/news/briefing/21677228-technology-behind-bitcoin-lets-people-who-do-not-know-or-trust-each-other-build-dependable. Accessed 31 October 2015.

European Commission. (2022). Blockchain Standards. https://digital.strateg.ec.europa.eu/en/policies/blockchain-standards.

Fairfield, J.A. (2014). Smart Contracts, Bitcoin Bots and Consumer Protection. *Washington and Lee Law Review Online* 7(12), 35–50.

Fennel, J. (2019). Understanding the Risks Associated with Blockchain. https://focus.world.exchanges.org/articles/understanding-risks-associated-blockchain. Accessed 25 July 2022.

Frank Cerveny, T.K. (2019). Research Announcement: Moody's Blockchain Standardization Will Amplify Benefits for Securitisations. www.moody's.com/research/Moodys-Blockchain-standardization-will-amplify-benefits-for-securitisation-PBS_1193318?Stop_mobi=yes&showPdf=true. Accessed 19 April 2020.

Ghosh, E., and Das, B. (2020). A Study on the Issue of Blockchain: Energy Consumption. *Proceedings of International Ethical Hacking Conference* 2019, 63–75. https://doi.org/10.1007/978-981-15-0361-0_5.

Glover, D.G., and Hermans, J. (2017). Improving the Traceability of the Clinical Trial Supply Chain. *Applied Clinical Trails* 26(12), 36–38.

Green, A. (2017). Will Blockchain Accelerate Trade Flows? *Financial Times*. www.ft.com/content/a36399fa-a927-11e7-ab66-21cc87a2edde. Accessed 9 November 2017.

Hackett, R. (2017b). Maersk and Microsoft Rested a Blockchain for Shipping Insurance. *Fortune*. http://fortune.como/2017/09/05/maersk-blockchain-insurance/. Accessed 6 September 2017.

Hasonova, H., Ui-Jun, B., Mu-gon, S., Kyunghee, C., and Myung-Sup, K. (2019). A Survey on Blockchain Cybersecurity Vulnerabilities and Possible Countermeasures. *The International Journal of Network Management* 29(2), 1–36. https://onlinelibrary.wiley.com/doi/epdf/10.1002/nem.2060?saml_referrer.

Higgins, S. (2017). The US Treasury is Testing Distributed Ledger Asset Tracking Coindesk. www.coindesk.com/us-treasury-testing-distributed-ledger-asset-tracking/. Accessed 3 October 2017.

Hurd, J., and Isaak, J. (2008). It Standardization: The Billion Dollar Strategy. In: *Standardization Research in Information Technology: New Perspective*. Aachen: IGI Global, Aachen University, pp. 20–26.

Institute, E.P.R. (2019). *Program on Technology Innovations: Blockchain—U.S. and European Utility Insights Market Intelligence Briefing Report*. Palo Alto, CA: Technical Report; Washington, DC: Electric Power Research Institute.

Iqbal, M. (2020). *A Reference Model for Security Risk Management of the Blockchain-Based Applications*. Tartu, Estonia: Institute of Computer Science, University of Tartu, Vol. 2613, pp. 1–8. https://ceur-ws.org/Vol-2613/paper5.pdf.

Iredale, G. (2021). Blockchain Risks Every CIO Should Know. https://101blockchains.com/blockchain-risks?sfw=paws1658794024.

Kabanda, G. (2021). Cybersecurity Risk Management Plan for a Blockchain Application Model. *Transactions on Engineering and Computer Science* 2(1), 1–18.

Kaplan, Z. (2022). What Are Problem-Solving Skills? Definition and Examples. www.theforage.com/blog/skills/problem-solving-skills#:~:text=Problem%2Dsolving%20 skills%20are%20the,and%20implement%20the%20best%20solutions. Accessed 24 October 2022.

König, L., Korobeinikova, Y., Ijoa, S., and Kieseberg, P. (2020). Comparing Blockchain Standards and Recommendations. *Future Internet* 12, 1–17. https://doi.org/10.3390/fi12120222.

Kshetri, N. (2018). Blockchain's Roles in Meeting Key Supply Chain Management Objectives. *International Journal of Information Management* 39, 80–89.

Lai, L., Zhou, T., Cai, Z., Liang, Z., and Bai, H. (2021). A Survey on Security Threats and Solutions of Bitcoin. *Journal of Cyber Security* 3(1), 29–44. https://doi.org/10.32604/jcs.2021.016349.

Ma, S., Wang, H., Dai, H.N., Cheng, S., Yi, R., and Wang, T. (2018). A Blockchain-Based Risk and Information System Control Framework. *IEEE Computer Society* 106–113. http://hdl.handle.net/11250/2594709.

Mahmood, Z., Arum, K.C., Rana, E., and Iftikhar, W. (2020). A Study on Issues and Challenges of Blockchain Technology in Malaysian Higher Education Institution. *International Journal of Psychological Rehabilitation* 24(5).

Malhota, A., O'Neill, H., and Stowell, D. (2022). Thinking Strategically About Blockchain Adoption and Risk Mitigation. *Business Horizons* 65(2), 159–171.

Meyers, M., and Rogers, M. (2004). Computer Forensics: The Need for Standardization and Certification. *International Journal of Digital Evidence* 3, 1–11.

Miah, M.S.U., Rahman, M., Hossain, S., and Al-Ahsan, A. (2019). Introduction to Blockchain. In: *Blockchain for Data Analytics*. Cambridge: Cambridge Scholars Publishing. https://www.researchgate.net/publication/343601688_Introduction_to_Blockchain.

Min, H. (2019). Blockchain Technology for Enhancing Supply Chain Resilience. *Business Horizons* 62(1), 35–45. https://doi.org/10.1016/j.bushor.2018.08.012.

Misra, K.B. (2016). Risk Analysis and Management: An Introduction. *Researchgate* 662–674. https://doi.org/10.1007/978-1-84800-131-2_41.

Morganti, G., Schiavone, E., and Bondavalli, A. (2018). Risk Assessment of Blockchain Technology. *2018 Eighth Latin-American Symposium on Dependable Computing (LADC)*, pp. 87–96. https://doi.org/10.1109/LADC.2018.00019.

Nakasumi, M. (2017). Information Sharing for Supply Chain Management Based on Blockchain Technology. *2017 IEEE 19th Conference on Business Informatics* 1, 140–149.

Nwulu, N., and Damisa, U. (2023). Chapter 1: Introduction to Industry 4.0 Technologies. In: *Energy 4.0, Concept and Applications*. https://doi.org/10.1063/9780735425163_001.

Nzuva, S. (2019). Smart Contracts Implementation, Applications, Benefits, and Limitations. *Journal of Information Engineering and Applications* 9, 63–75. https://doi.org/10.7176/JIEA/9-5-07

Onwutalobi, A.-C. (2011). Overview of Cryptography. *SSRN Electronic Journal*. https://doi.org/10.2139/ssrn.2741776.

Patibandla, R.S.M.L., and Vejendla, L.N. (2022). Significance of Blockchain Technologies in Industry. In: Baalamurugan, K., Kumar, S.R., Kumar, A., Kumar, V., and Padmanaban, S. (eds) *Blockchain Security in Cloud Computing. EAI/Springer Innovations in Communication and Computing*. Cham: Springer. https://doi.org/10.1007/978-3-030-70501-5_2.

Sahu, M. (2022). Skills Needed to Become a Blockchain Developer. www.upgrad.com/blog/skills-needed-to-become-blockchain-developer/. Accessed 24 October 2022.

Suman, A.K., and Patel, M. (2021). An Introduction to Blockchain Technology and Its Application in Libraries. *Library Philosophy and Practice (e-journal)* 1–12.

Swan, M. (2015). *Blockchain: Blueprint for a New Economy*. Sebastopol, CA: O'Reilly Media.

Tapscott, D., and Tapscott, A. (2017). How Blockchain Will Change Organizations. *MIT Swan Management Review* 58(2), 10–13.

Tomaszewski, M. (2022). Top 8 Critical Thinking Skills and Ways to Improve Them. https://zety.com/blog/critical-thinking-skills#:~:text=The%20key%20critical%20thinking%20skills,mindedness%2C%20and%20problem%2Dsolving. Accessed 24 October 2022.

Tucci, L. (2022). What is Risk Management and Why Is It Important? www.techtarget.com/searchsecurity/definition/what-is-risk-management-and-why-is-it-important. Accessed 26 July 2022.

Wang, H., Mylopoulos, J., and Liao, S. (2002). Intelligent Agents and Financial Risk Monitoring Systems. *Communication of the ACM* 45(3), 83–88.

Wang, S., and Jean-Phillippe, V. (2017). Buzz Factor or Innovation Potential: What Explains Cryptocurrencies Returns? *PLoS One* 12(1). https://doi.org/10.1371/journal.pone.0169556.

White, B.S., King, C.G., and Holladay, J. (2020). Blockchain Security Risk Assessment and the Auditor. *Journal of Corporate Accounting & Finance* 31(2), 47–53.

White, G.R. (2017). Future Applications of Blockchain in Business and Management: A Delphi Study. *Strategic Change* 26(5), 439–451.

Willis Towers Watson. (2019). Cryptocurrency: Risk Management Overview. www.willytowerswatson.com. Accessed 26 July 2022.

Yu, L., Zheng, X., and Biao, Y. (2019). Research on Data Structure Course Teaching System Based on Open Teaching Model. *Advances in Social Science, Education and Humanities Research* 268, 441–450.

Zamani, E., He, Y., and Phillips, M. (2018). On the Security Risks of the Blockchain. *Journal of Computer Information Systems* 60(6), 495–506. https://doi.org/10.1080/08874417.2018.1538709.

Zhao, J.L., Fan, S., and Yan, J. (2016). Overview of Business Innovations and Research Opportunities in Blockchain and Introduction to the Special Issue. *Financial Innovation*, 2(28). https://doi.org/10.1186/s40854-016-0049-2.

Zheng, Z., Xie, S., Dai, H.N., Chen, X., and Wang, H. (2015). Blockchain Challenges and Opportunities: A Survey. *International Journal of Web and Grid Services* 14(4), 352–375.

7 Cloud Computing

7.1 INTRODUCTION TO AND COMPONENTS OF CLOUD COMPUTING

In recent years, cloud computing has developed rapidly, evolving into a pivotal component of an organization's information technology strategy. This evolution has played a crucial role in freeing numerous technological solutions from the constraints of outdated software and hardware licensing models, transforming data center paradigms, fostering openness, reinvention, and, to some extent, democratizing the delivery of IT services. This transformative shift has substantially enhanced users' access to data, applications, and business services. Recognized as the fifth generation of architecture in the IT industry, cloud computing is heralded for its implications in providing greater flexibility and availability at a reduced cost. Regarded as a revolution in information technology, cloud computing has gained considerable momentum and significance (Khan 2019; Jaiswal 2017; Kilari 2018; Sedani and Doshi 2015).

The term "cloud computing" is designed for network access on-demand permits sharing a group of reconfigurable computing resources that can be utilized extensively and abandoned with little management or service provider intervention. Cloud computing has advancing more quickly than ever due to organizations of all kinds adopting new technology. According to industry analysts, this trend will only continue to expand and intensify in the upcoming few years (Madaan et al. 2015).

The term "cloud computing" refers to users storing or moving their data to a "cloud" and accessing it in various ways. Cloud computing is where flexible and adaptive information technology-enabled capabilities are offered as a service to multiple outside consumers through Internet developments (Kilari 2018). It is a sort of Internet-based computing that gives pooled processing resources and data to computers and other devices on demand (Samreen et al. 2018; David et al. 2022; Nwulu and Damisa 2023). It refers to Internet-based computing where various services, such as servers, storage, and applications, are supplied to an organization's computers and devices through the Internet instead of using local servers or personal devices to manage applications.

Cloud computing focuses on sharing the available computer resources, making shared data, software, and resources available to computers and other devices as needed, similar to the electricity grid. It is the culmination of numerous prior attempts at large-scale computing since it offers seamless access to essentially infinite resources. It provides business owners with top-notch amenities and adaptable infrastructure (Haghighat, Zonouz, and Abdel-Mottaleb 2015).

The term "cloud" is a metaphor for "the Internet" in this context; hence, its services are tremendously useful in the personal and professional life of an individual because they may be used in any enterprises. The most extensive example

DOI: 10.1201/9781003522102-9

is in Internet business apps, as shown in Figure 7.1 (Samreen et al. 2018). Using this architecture, one can join a network from any location to access information and computer resources. When an individual creates a document using application software like Notepad, Microsoft Word, and so forth installed on a computer and stores the document in a mailbox, the individual can view the same data online if Internet availability is available (Samreen et al. 2018). When cloud computing is used, data is stored online rather than on a laptop or personal computer. Cloud computing also entails delivering computing resources over the Internet. Thus, instead of storing data on hard drives, it can be easily stored on the Internet, which makes lives better due to the efficiency, availability, ease of access, and remote access of information (Madaan et al. 2015; Ashraf et al. 2013). Various alternatives for storing and processing data in external data centers are available to users and enterprises owing to cloud computing and storage solutions. Resource sharing is required for coherence and scale economies throughout a network. The advantages of cloud computing, such as its low service prices, high performance, scalability, accessibility, and availability, have contributed to its popularity.

John McCarthy developed the idea of cloud computing in the 1960s, stating that computation might someday be organized as a public utility. Consequently,

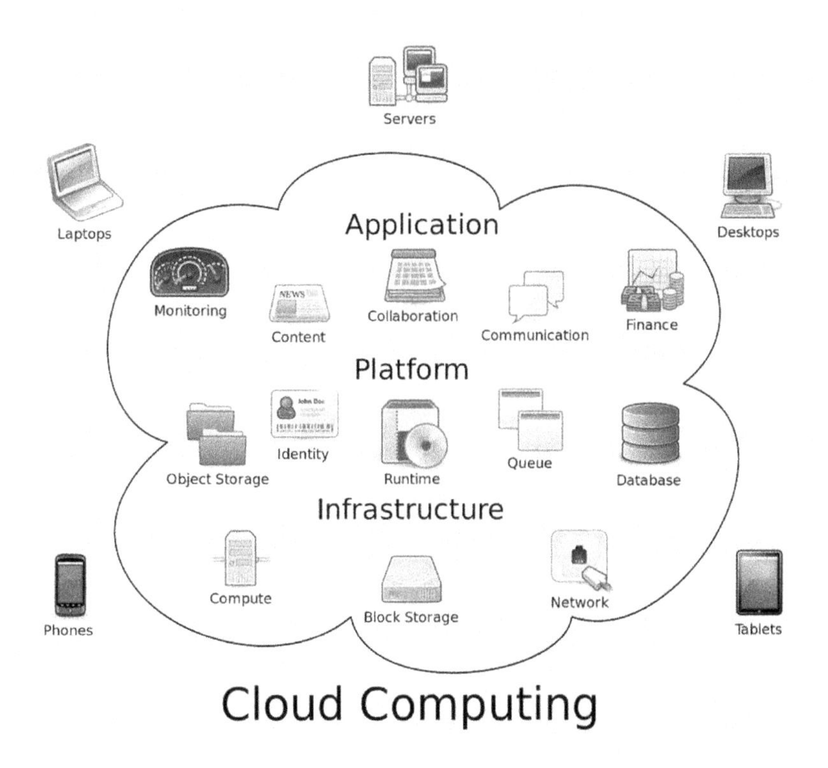

FIGURE 7.1 Cloud computing utilization (Samreen et al. 2018, p. 785)

the first time the characteristics of cloud computing were examined was in 1966 by Douglas Parkhill in his book *The Challenge of the Computer Utility*. The term "cloud" has its roots in the telecommunications industry, where telecom companies began to offer virtual private network services with comparable quality of service at a much lower cost. Before developing VPNs, they offered specialized point-to-point data lines, which are little more than bandwidth wasters. However, they could switch traffic to balance network usage by employing VPN services. This is now expanded by cloud computing to include servers and network infrastructure. Industry players have embraced and adopted cloud computing in large numbers. As an illustration, Amazon significantly influenced the creation of Amazon Web Service in 2006. Along with this, research initiatives in cloud computing were also launched by Google and IBM, with the first open-source private cloud deployment platform called Eucalyptus.

Also, cloud computing components refer to platforms like the front end, back end, cloud-dependent delivery, and the used network necessary for the smooth operations of cloud computing. Clients interact with the front end, including client-side interfaces and programs necessary for accessing cloud computing platforms (Rebah and Bensta 2018). Tablets, smartphones, thin and fat clients, web servers, and other mobile devices make up the front end. Web servers include Chrome, Firefox, Internet Explorer, and others; while utilizing the back end, the service provider manages all the resources required to provide cloud computing services. It includes massive data storage, safety precautions, virtual machines, servers, traffic management systems, and so forth. A network, typically the Internet connection, connects the front end and back end to other devices (Jaiswal 2017). The essential components of cloud computing, as shown in Figure 7.2, include the following:

Client infrastructure: The front end includes the client infrastructure. For cloud interaction, it offers a graphical user interface.

Application: Any software or platform a client wants to access can be an application.

Service: A cloud service that controls the service you access based on the customer's needs. Three different services are available with cloud computing include the following:

i. **Software as a service (SaaS):** It is sometimes referred to as cloud application services. Most SaaS programs run immediately through the web browser, so we don't need to download and install them. Examples include Salesforce and Google Apps. Dropbox, Slack, Hubspot, and Cisco WebEx (Jaiswal 2017).

ii. **Platform as a Service (PaaS):** These services are also referred to as cloud platforms. SaaS and PaaS are relatively similar; however, SaaS offers software access through the Internet without requiring a platform, whereas PaaS provides a platform for software creation. For instance, consider OpenShift, Magento Commerce Cloud, Windows Azure, and Force.com (Rebah and Bensta 2018).

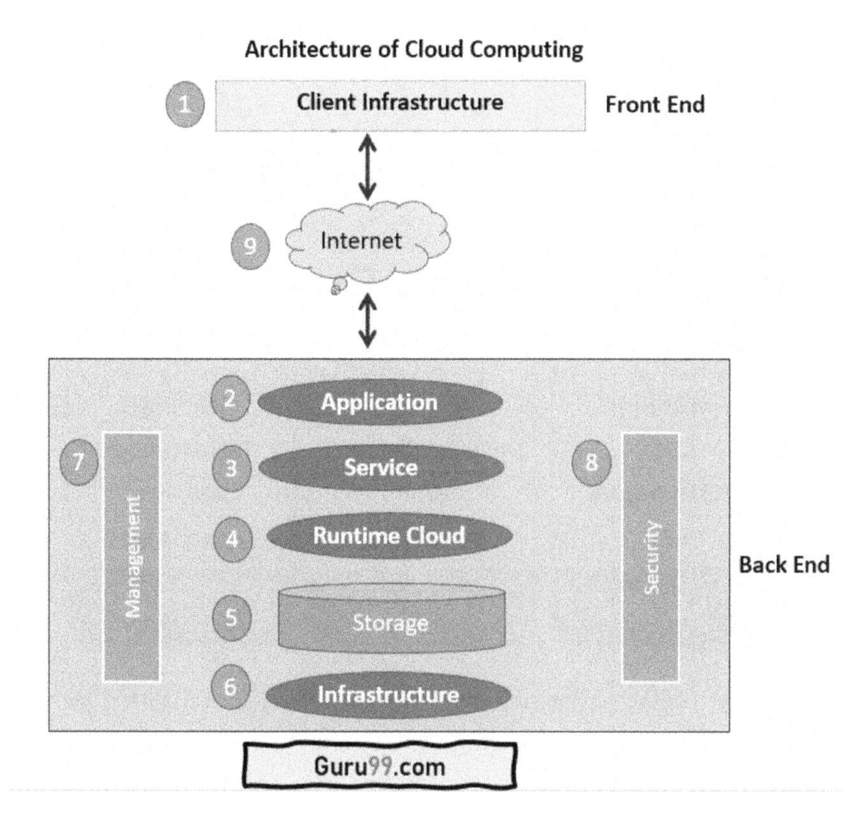

FIGURE 7.2 Components of cloud computing (Peterson 2024, para 6)

iii. **Infrastructure as a service (IaaS):** It is sometimes referred to as cloud infrastructure services, as exemplified in Figure 7.3. It oversees runtime environments, middleware, and data for applications. Examples include Amazon Web Services EC2, Google Compute Engine (GCE), and Cisco Metapod (Rebah and Bensta 2018).

Runtime cloud: Runtime cloud offers virtual machine execution and runtime environments.

Storage: One of the most crucial elements of cloud computing is storage. It offers a sizable quantity of cloud storage space for managing and storing data.

Infrastructure: It offers network-wide, host-level, and application-level services. The hardware and software components that make up cloud infrastructure and are essential to the operation of the cloud computing model include servers, storage, network devices, virtualization software, and other storage resources.

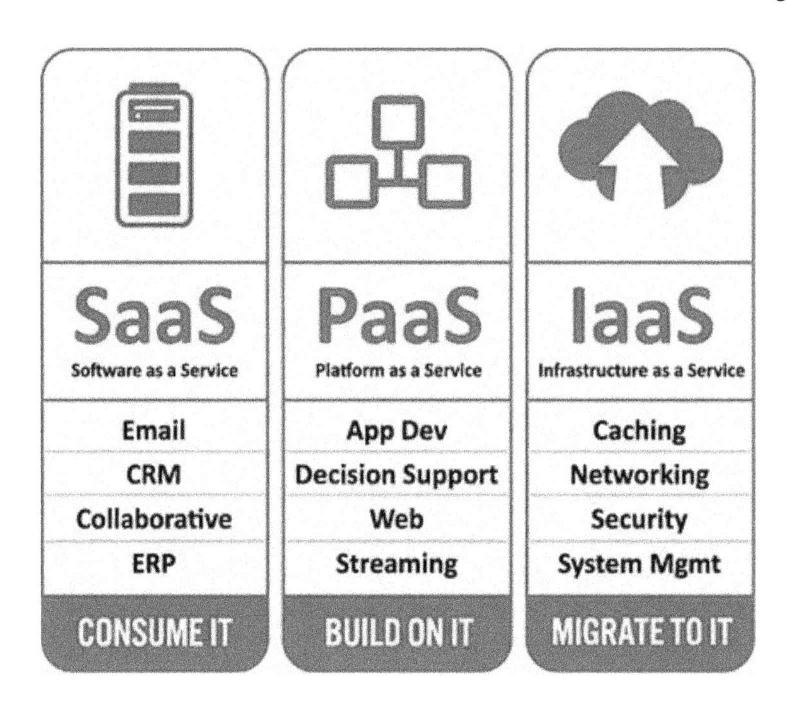

FIGURE 7.3 Cloud computing infrastructure services (Samreen et al. 2018, p. 786)

Management: Management is used to build coordination among teams and developers on components such as applications, services, runtime clouds, storage, infrastructure, and other security issues in the background.

Security: Security is a built-in element of the back end of cloud computing. At the back end, a security mechanism is implemented.

Internet: The Internet serves as a conduit for front—and back-end interaction and communication.

7.2 RISKS ASSOCIATED WITH CLOUD COMPUTING

Risk is defined as the likelihood that a given danger may take advantage of weaknesses in one or more assets, harming an organization according to ISO/CEI 13335-1:2009. It is quantified in terms of the likelihood of an event and its repercussions (Rebah and Bensta 2018). Bassett (2015) highlighted nine risks associated with cloud computing: interoperability, portability, and data lock-in; compliance; legality and auditability; security; accessible data; incident response, notice, and remediation; virtualization; governance; and corporate risk management. Risks associated with cloud computing include the following:

Compliance risk: Compliance, according to the Cloud Security Alliance (2013), is the knowledge of and observance of duties (such as corporate social responsibility, applicable laws, and ethical norms), as well as the evaluation and prioritization of remedial actions deemed required and suitable. Preexisting compliance and information security standards may not be appropriate because they were not initially created with cloud computing in mind, making compliance a significant barrier to cloud computing. An organization must know the rules and laws that apply to that particular region if it is considering using a cloud service provider from another country. Compliance issues would also be impacted by the information stored in the cloud (Queensland Government 2013).

Legality and auditability risk: This addresses an organization's compliance with the law and, if inspected, being held accountable. Organizations must abide by local laws and regulations. Data can, however, be hosted abroad in a variety of countries and regions as a result of cloud computing. This has important legal implications because local laws govern data stored there in those nations. The benefits of cloud computing may be compromised by hosting restrictions and service-level agreements. The ability of cloud computing to host data in numerous places worldwide is one of its distinctive features. However, information stored in the cloud is subject to the laws of the nation where the data is physically located, which could be an issue.

Security risk: This is possibly the most significant risk cloud computing faces as a 4IR technology. The cloud computing provider is responsible for ensuring customer data is secure. The client must ensure that the supplier can offer the required level of security for their data. Data must be dependable, secure, legitimate, and secret. Regarding cloud computing, particular security threats include identity authentication and access management, human factors, surface attacks and vulnerabilities, security as a service, application security, traditional security, business continuity, disaster recovery and encryption, and key management.

Accessibility risks: Bassett (2015) has referred to the availability of data everywhere and anywhere as "everywhere accessible data" and highlighted the issue regarding mobile devices and collaboration tools. More and more mobile devices, including cell phones or tablets (such as the iPad and Surface tablet), are being used for mobile access due to the growth in cloud computing. According to Gartner prediction, at least 60% of information workers will use a content application via a mobile device (Lavenda 2013). Risk occurs when businesses are convinced to move their data to the cloud due to the rapid expansion of mobile solutions and productivity tools. Frequently, there may be insufficient security or compliance difficulties (Buckley 2013). It is also possible to dispute the practicality of cloud computing for records management.

Incident response, notification, and remediation risk: This risk is related to the occurrence of an incident and how it is handled, both of which can be risky. An incident is an event that could have either a favorable or unfavorable impact on the entity. An incident's potential for harm could be raised if it is handled improperly or slowly. Businesses, in particular, should be informed of how the cloud service provider handles incidents before using cloud computing. When there is an incident, the service provider must inform the clients and let them know how it is handled. Although a new incident response procedure does not need to be designed for cloud computing, the organization must modify its current incident response procedures to consider the new cloud environment. The incident response plan must be formalized and recorded and include a list of everyone's responsibilities.

Virtualization risk: In the context of cloud computing, virtualization is defined as a virtual, non-physical system where resources are gathered and shared. The virtual machine's security rests with the cloud client. However, the secured virtual machine images are the cloud service provider's responsibility (Bouayad et al. 2012; Sosinsky 2011). The following virtualization concerns are listed by The Cloud Security Alliance (2011): performance issues, virtual machine guest hardening, inter–virtual machine attacks and blind spots, operational complexity brought on by the sprawl of virtual machines, virtual machine encryption, instant-on gaps, virtual machine data destruction, virtual machine image tampering, and in-motion virtual machines are just a few of the issues that need to be addressed.

Governance and enterprise risk management: These components deal with the development and implementation of organizational procedures, frameworks, and guidelines that are used to uphold information security governance, compliance, and risk management (Cloud Security Alliance 2013). The processes, structures, and controls that regulate them can become ungoverned due to data mismanagement or network difficulties occasioned by espionage or intellectual rivalry, resulting in a loss of control over these problems. An opportunity for risk exists for potential cloud users. Failure of governance may make it more difficult for the organization to follow legal and regulatory requirements. It is necessary to demonstrate the organization's ability to show the validity, dependability, and integrity of the data it stores on the cloud. This is complicated further by cloud service providers that might not want to share usage and access logs with their users for auditability. Unfortunately, clients may find it equally challenging to use their monitoring tools for this. Organizations need to take these challenges into account. The traditional data center may be a more practical choice if the service provider is unable to exchange information, which could impair the auditability of the stored data.

Interoperability, portability, and data lock-in: The ability of all cloud computing components to trade with new and different components from multiple providers while still operating is known as interoperability. Regardless of the operating system, provider, location, infrastructure, or application program interface, portability describes the capacity of the applications' constituent parts to be relocated and reused in another area. However, data lock-in with a cloud service provider might result from poor interoperability and portability. This is because there aren't any standardized APIs or processes, which can make it expensive or very challenging for users to switch to another service provider because they are now locked into the development environment of the present cloud service provider.

Viability risk: The viability of cloud computing is thought to come with risks like hidden variable expenses, shared reputation, and accountability issues, along with their respective impacts. Hidden expenses can make what appears to be a workable option into an expensive choice when businesses are not fully aware of what they are being charged for under the pay-per-use cloud model. A client's instances may be idle for part of an hour when charging by the hour for cloud services. As businesses try to outsource their data to cut expenditure on IT, cloud computing has become more popular due to its low costs. The risk associated with hidden variable costs is a drawback of this techno-economic model. An organization's capacity to grow may depend on its workload, which may change at unforeseen times and result in unpredictably high costs when more resources are needed (Himmel 2012).

Availability and reliability risk: The service's dependability and availability are crucial advantages of cloud computing (Bassett 2015). These characteristics can, however, come with risks of their own. The service level commitments necessary for vital business processes are a significant challenge. It is frequently the case that these offers are not part of the cloud service provider's actual service. In such cases, the client must specify the service level requirements the cloud service provider requires and ensure there are consequences for the cloud service provider if these are not satisfied.

7.3 RISK MANAGEMENT TECHNIQUES IN CLOUD COMPUTING

Risk is commonly defined as the product of the magnitude of an unfavorable result multiplied by the likelihood that it will occur. In the context of cloud computing, system threats, exploitable weaknesses, and the potential outcomes of such exploits all contribute to the overall risk, typically within the cloud component. This book advocates for utilizing the Cloud Security Risk Management Framework (CSRMF) proposed by Youssef (2019). This framework serves as a method for identifying, describing, assessing, addressing, and monitoring security threats throughout the lifespan of cloud services (Youssef 2019).

The CSRMF provides a structured approach to navigate the intricacies of cloud security, offering a comprehensive framework for risk identification, assessment, and management throughout the cloud service lifecycle.

The objectives of the CSRMF include the following:

- Identifying risks that pose threats to the cloud
- Analyzing and evaluating the identified risks

- Implementing optimal treatment actions to reduce the likelihood and impact of these risks
- Regularly monitoring the currency of identified risks to ensure the validity of treatment actions
- Establishing a dynamic relationship between service providers, end users, and cloud solution providers during risk management to ensure compliance with service level agreements

7.3.1 Risk Identification

The pivotal phase in effective risk management is identifying security issues impacting organizations and businesses implementing cloud computing (CC). The efficiency and meaningfulness of the risk management approach greatly depend on accurately identifying and understanding these risks. Various factors such as the application area, project nature, available resources, regulatory requirements, and client objectives play a significant role in determining the most suitable risk identification method.

Risks in this context arise from security threats exploiting ambiguities in the CC platform, potentially causing harm to an organization's assets and impeding the achievement of its goals. Therefore, accurately identifying assets, weaknesses, and threats within the CC platform is crucial for accurate risk assessment. The CSRMF adopts a hybrid strategy that combines two risk identification techniques, acknowledging that no single scientific approach can guarantee the detection of all risks. CSRMF employs documented knowledge acquisition and brainstorming for effective risk identification (Youssef 2019).

Documented knowledge acquisition: This strategic approach involves a comprehensive gathering and meticulous study of the CC risk domain materials. This extensive collection includes various sources such as books, surveys, papers, and regulatory documents. Delving into the wealth of literature is a crucial step to discern and identify potential risks and threats within the CC landscape. A particular focus is given to authoritative publications, exemplified by the European Network and Information Security Agency publication, which serves as a valuable repository for insights into the intricate realm of CC risks. This method ensures a well-rounded and informed foundation, enriching the risk assessment process with knowledge gleaned from diverse and reputable sources.

Brainstorming: This process involves generating and evaluating ideas to address a cloud computing problem or scenario within the CSRMF. A collaborative effort is initiated, including a brainstorming team comprising information security specialists and various organizational stakeholders. This collective effort aims to comprehensively assess the organization's assets, weaknesses, and potential risks. The CSRMF utilizes a series of group workshops to identify risks, enabling a thorough exploration of issues and the application of innovative risk identification methods. The group employs this insight to pinpoint various risks based on the knowledge acquired during documented knowledge acquisition. The identified risks undergo scrutiny by an impartial stakeholders' group. If satisfaction is achieved, the risk management process seamlessly progresses to the next stage; however, if not, the risk identification phase is revisited. This systematic and collaborative approach within the CSRMF ensures a robust examination of potential risks and encourages innovative thinking to address challenges in cloud computing scenarios effectively.

Risk analysis: Within the risk analysis process, risk likelihoods and implications are evaluated, and the CSRMF employs a quantitative method for this analysis. The Delphi method, a widely utilized consensus-based approach, is pivotal in this process. The Delphi method incorporates three

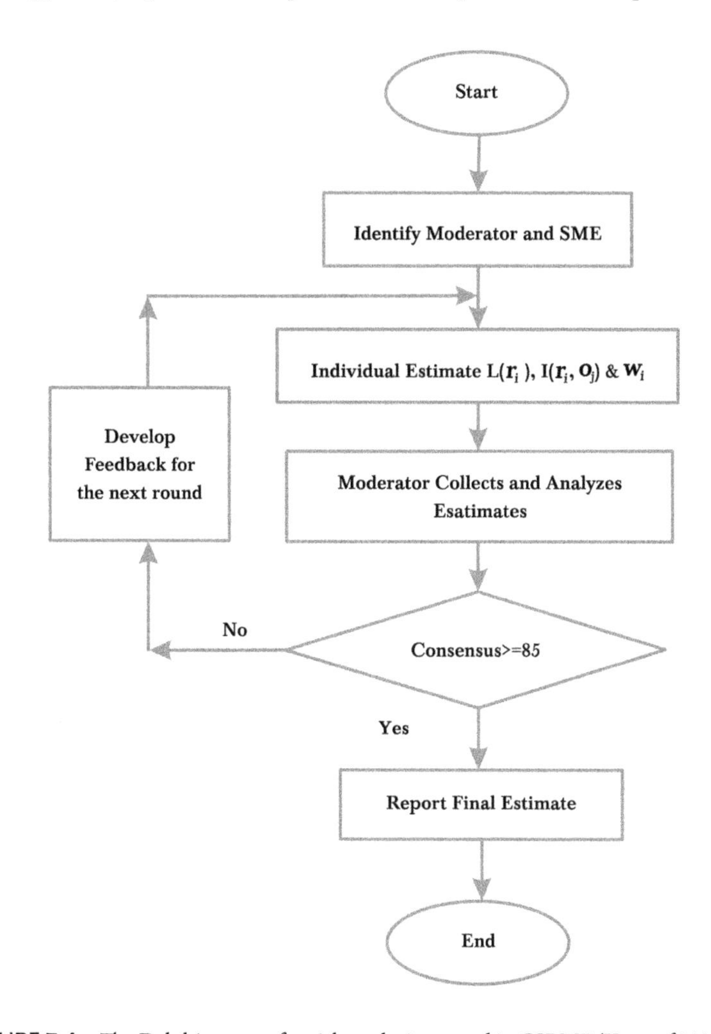

FIGURE 7.4 The Delphi process for risk analysis as used in CSRMF (Youssef 2019, p. 190)

essential components, as illustrated in Figure 7.4: a structured and iterative information flow, participant anonymity to ease peer pressure and performance apprehension, and iterative participant feedback until consensus is realized. In the CSRMF Delphi technique, a moderator plays a crucial role, overseeing and facilitating information gathering from a carefully selected group of subject matter experts (SMEs). These SMEs possess in-depth knowledge regarding the probabilities and effects of risks specific to the organization's particular line of business. The structured and iterative flow, participant anonymity, and continuous feedback until consensus ensure a robust and comprehensive risk analysis within the CSRMF framework.

Risk evaluation: Risk analysis involves evaluating the risk level, often referred to as risk severity, to ascertain whether the organization can tolerate or accept the risk. A pertinent committee comprising experts and stakeholders must collaboratively formulate, endorse, and document the acceptable risk criteria. If the evaluated risk level surpasses the acceptable threshold, the specific risk necessitates treatment or the implementation of enhanced defenses. If a risk index (ri) falls below the predefined threshold, it may be deemed acceptable; otherwise, it warrants rejection or avoidance.

Risk response: Effectively addressing any unacceptable risk involves lowering its risk level until it falls below the predefined threshold. The objective of risk response is to formulate viable and cost-effective options for managing risks that are deemed unacceptable. Several response strategies are available, including risk avoidance, risk transfer, or sharing with a third party. Additionally, risk mitigation (reduction) is employed, which involves managing the likelihood of the risk occurring or the impact of its consequences. By implementing these response measures, organizations aim to maintain a risk profile that aligns with their risk tolerance and strategic objectives.

Risk monitoring and control: The conclusive phase of the CSRMF involves vigilant monitoring and thorough assessment of the effectiveness of the prescribed risk remedies and ongoing control measures. To facilitate this, the Delphi approach can be employed to provide risk control measures and evaluate the extent of risk level reduction. This iterative process ensures that the chosen risk response strategies are continually aligned with the evolving risk landscape, allowing organizations to dynamically adapt their risk management approach. Regular evaluations enable timely adjustments to the risk mitigation measures, ensuring a proactive and adaptive risk management framework that stays robust in the face of changing circumstances.

7.4 RISK MANAGEMENT SKILLS FOR CLOUD COMPUTING

Programming skills: Popular programming languages work better with cloud-based technologies. These include established ones like Java, JavaScript, and Python and newer ones like Go and Scala. SQL, NoSQL, and Linux knowledge are prerequisites for database programming. Cloud architects and administrators occasionally need to create code as well; therefore, this is crucial for all software developers.

Data analysis skills: Risk managers need analytical abilities to gather and use that data to make critical decisions. They must also look for flaws and gaps in the infrastructure, systems, and other places that may have gone unnoticed by others. Risk managers must understand the effective use of enterprise risk planning systems while managing enormous amounts of data.

Artificial intelligence skill: A basic understanding of AI and associated technologies is required for any IT engineer who wants to work in the cloud. Machine learning algorithms make the production of insights based on massive data sets possible. The cloud is being developed with a lot of apps that don't need human supervision. For instance, chatbots and virtual assistants reply to inquiries and requests after swiftly evaluating and deciphering user input. Business intelligence and intelligent Internet of Things (IoT) devices are a part of the complex web of networks that cloud developers must oversee.

Project management skills: Knowing risk management, service contracts, legal issues, and connections with other operations are crucial. Risk managers will be better equipped to recognize any potential obstacles hindering their team's capacity to generate results if they have a solid understanding of project management and its objectives.

Database management skills: Database expertise that goes beyond what is typically required in the data center is needed to work as a cloud engineer. These include database cloning, size restrictions, storage performance, and multi-cloud activities. Databases swiftly migrate to the cloud like many other IT functional areas. Cloud databases can be spread out across a cloud architecture, as opposed to traditional databases, which are housed in data centers and administered in specific geographic areas. Cloud providers offer a specific type of cloud database to store and manage customer data called database-as-a-service (DBaaS). Although SQL is the preferred database language for the cloud, NoSQL is gaining popularity as a viable alternative to SQL's inflexible design.

Change management and adaptability skills: Change management is the administration of "moves and changes" inside an IT infrastructure. It entails developing a comprehensive plan that addresses every part of a change to IT resources, whether a server, network connection, or database. Failure to implement a change of service is one of the leading causes of Internet service interruptions. They frequently have late-night maintenance windows scheduled for them. Every intended change should be supported by a written method of procedure, and there should be a rollback strategy in case issues develop. This strategy for service enhancements has been widely employed in conventional IT systems and needs to be used for cloud migrations.

Strategic thinking skills: The ability to view the big picture is necessary, but one also need to be able to dissect it to uncover opportunities and answers. With the knowledge that risks have the potential to influence competitors and alter the nature of business differentially, one can adapt risk management protocols to strategic difficulties by combining a strategic management focus with a multidimensional model of strategic risk.

Decision-making skills: Risk management entails making choices that will affect risk in a foreseen and managed manner. Risk managers need to be capable of making wise decisions. To make the best decision possible, people must be capable of obtaining information, weighing their options, and making a final decision.

Problem-solving skills: Risk management is a strategic business. At the upper levels, an individual can develop strategies and procedures for risk management throughout an entire organization, which will require a methodological yet original approach. If potential hazards are identified, you must be curious to examine the company's problem and arrive at the key solutions. If you desire to solve problems, you'll be able to push for accurate solutions and search strategically for them.

Leadership and management skills: Risk managers require leadership abilities to motivate and guide workers. Managers need the team's respect to overcome problems in risk management, which may mean disrupting the status quo. Solid management and leadership abilities are required to support people in managing risk in their domain, to help the team accept changes, and to make it all happen. Good leaders can command respect while maintaining a cordial relationship with the entire team.

Financial intelligence skills: Every firm faces some financial risk; hence, a competent manager needs to recognize the risk, quantify it, and implement mitigation measures. A risk manager can concentrate on contingency planning and prepare for the unexpected if they can distinguish between risks that can be handled and those that are out of control.

Business understanding skills: Understanding how a business operates and all the various internal and external elements that can impact its performance is necessary for identifying and estimating the risks to a firm.

Communication skills: Making sure that everyone is aware of any significant risks and the organization's risk management strategy is crucial. This calls for effective communication with all target groups, including the board of directors and individual employees. Risk managers must be able to establish connections across all chains, and not just with their direct reports. Modifications to the business or team-wide learning will inevitably need to be made as you identify and manage risk. It's your responsibility to convince everyone why listening to your advice and lessons is crucial. The risk management approach will only be successful if all parties are on board.

7.5 SUMMARY

Cloud computing supplies computer resources such as storage, processing power, databases, networking, analytics, artificial intelligence, and software applications over the Internet. The need for mobile computing gave rise to the concept of cloud computing, which is now becoming more popular in the IT industry. It provides the user access to information, programs, and storage that are not kept on their computer.

This chapter has covered the various risks associated with cloud computing, the techniques for managing the risks, and the skills needed to optimize the management of the risks. This is because most of the risks are cloud related. Thus, a cloud provider must therefore manage the risks connected to the cloud computing environment using the cloud computing risk management framework to recognize, evaluate, and prioritize those risks to lower those risks, enhance security, increase trust in cloud services, and alleviate organization's worry about using a cloud environment.

REFERENCES

Ashraf, A., Hartikainen, M., Hassan, U., Heljanko, K., Lilius, J., Mikkonen, T., Porres, I., Syeed, M., and Tarkoma, S. (2013). Introduction to Cloud Computing Technologies. https://doi.org/10.13140/2.1.1747.8082.

Bairagi, S., and Bang, D. (2015). Cloud Computing: History, Architecture, Security Issues. *International Journal of Advent Research in Computer and Electronics (IJARCE). Special Issue, National Conference—Convergence 2015.* https://www.researchgate.net/profile/Ankur-Bang/publication/323967455_Cloud_Computing_History_Architecture_Security_Issues/links/5ab52b4aa6fdcc46d3b2aa66/Cloud-Computing-History-Architecture-Security-Issues.pdf

Bassett, C. (2015). *Cloud Computing and Innovation: Its Viability, Benefits, Challenges and Records Management Capabilities.* Master's Dissertation. University of South Africa, Pretoria.

Bouayad, A., Blilat, A., El Houda Mejhed, N., and Ghazi, M. (2012). *Cloud Computing: Security Challenge in Institute of Electrical and Electronics Engineers.* Fez, Morocco: Colloquium in Information Science and Technology (CIST), 22–24 October.

Buckley, C. (2013). The Cloud: Mitigating Risks as You Relinquish Control. *Tech Republic.* www.techrepublic.com/blog/tech-decision-maker/the-cloud-mitigating-risks-asyou-relinquish-control.

Cloud Security Alliance. (2013). About. https://cloudsecurityalliance.org/about/.

David, L.O., Nwulu, N.I., Aigbavboa, C.O., and Adepoju, O.O. (2022). Integrating Fourth Industrial Revolution (4IR) Technologies into the Water, Energy & Food Nexus for Sustainable Security: A Bibliometric Analysis. *Journal of Cleaner Production* 363. https://doi.org/10.1016/j.jclepro.2022.132522.

Haghighat, M., Zonouz, S., and Abdel-Mottaleb, M. (2015). CloudID: Trustworthy Cloud-Based and Cross-Enterprise Biometric Identification. *Expert Systems with Applications* 42(21), 7905–7916. http://dx.doi.org/10.1016/j.eswa.2015.06.025.

Himmel, M.A. (2012). *Qualitative Analysis of Cloud Computing Risks and Framework for the Rationalization and Mitigation of Cloud Risks.* PhD Thesis. Pace University, New York.

Jaiswal, M. (2017). Cloud Computing and Infrastructure. *SSRN Electronic Journal* 4, 742–746.

Khan, S. (2019). Cloud Computing: Issues and Risks of Embracing the Cloud in a Business Environment. *International Journal of Education and Management Engineering* 9, 44–56. https://doi.org/10.5815/ijeme.2019.04.05.

Kilari, N. (2018). Cloud Computing—An Overview & Evolution. *International Journal of Scientific Research in Computer Science, Engineering and Information Technology* 3(1), 149–152.

Lavenda, D. (2013). Are You Ready for the New Mobile Reality. *Fast Company.* https://www.fastcompany.com/3009617/are-you-ready-for-the-new-mobile-reality.

Madaan, P., Singh, J., Kumar, M., and Kumar, S. (2015). Introduction to Cloud Computing. *International Journal of Computers & Technology* 14, 6097–6101. https://doi.org/10.24297/ijct.v14i9.1853.

Nwulu, N., and Damisa, U. (2023). Chapter 1: Introduction to Industry 4.0 Technologies. In: *Energy 4.0, Concept and Applications.* https://doi.org/10.1063/9780735425163_001.

Peterson, R. (2024). Cloud Computing Architecture and Components. *Guru99.* www.guru99.com/architecture-of-cloud-computing.html.

Queensland Government. (2013). Risks of Cloud Computing. www.business.qld.gov.au/business/running/technology-for-business/cloud- computing-business/cloud-computing-risks.

Rebah, H., and Bensta, H. (2018). *Cloud Computing: Potential Risks and Security Approaches*, pp. 69–78. https://doi.org/10.1007/978-3-319-66742-3_7.

Samreen, S.N., Khatri-Valmik, N., Salve, S.M., and Khan, P.N. (2018). Introduction to Cloud Computing. *International Research Journal of Engineering and Technology* 5(2), 785–788. www.irjet.net/archives/V5/i2/IRJET-V5I2174.pdf.

Sedani, J., and Doshi, M. (2015). Cloud Computing: From the Era of Beginning to Present. *International Journal of Novel Research in Computer Science and Software Engineering* 2(2), 33–38.

Sosinsky, B. (2011). *Cloud Computing Bible.* Indianapolis: Wiley Publishing.

Youssef, A. (2019). A Framework for Cloud Security Risk Management Based on the Business Objectives of Organizations. *International Journal of Advanced Computer Science and Applications.* http://dx.doi.org/10.14569/IJACSA.2019.0101226.

8 Robotic Technology

8.1 INTRODUCTION TO AND COMPONENTS OF ROBOTICS

The application of robotics spans various domains, integrating disciplines such as mechanical engineering, electrical engineering, computer science, electronics, sensors, actuators, and artificial intelligence. This multidisciplinary nature necessitates a profound understanding of matrices, vectors, derivatives, physics fundamentals, integrals, servo motor selection and design, gear selection, and correction techniques for comprehending the intricate mechanical design of robots. However, programming serves as the primary mode of communication with robots. Hence, a comprehensive grasp of hydraulics, pneumatics, and programmable logic systems and the ability to establish effective communication between sensors and actuators is essential (Sharakhatreh 2011).

The term "robotics" was coined by Czech author Karel Čapek in his 1920 play *R.U.R* (Rossum's Universal Robots). Originating from the Slavic word "robota," which translates to "work" or "job," robots were conceptualized by his brother Josef Čapek (Asimov 2014). Therefore, the study of robotics emerges as the science and technology of robots.

The inception of robotics brought forth fundamental laws, as articulated by Isaac Asimov in 2014. These laws dictate that robots should not harm humans, must obey human commands, and should not engage in self-destruction. These foundational principles guide the ethical development and application of robotics in various fields (Asimov 2014).

Moreover, robotics subcategories can be broadly classified into two main categories: ground robots and aircraft robots. Ground robots encompass a diverse range, including agents, automobiles, mobile robots, and smart home systems. On the other hand, aircraft robots contain miniature satellites, airships, unmanned aerial vehicles, and similar entities.

Further classification of robotics introduces the distinction between service robots and field robotics. Service robotics includes domestic and service-oriented robots, such as those found in households, while field robotics encompasses marine, airborne, and ground robots. With their versatility in various environments, ground robots can operate as mobile or static entities. The advent of sensors, edge devices, and other actuators has facilitated the creation of mobile robots designed for movement and the execution of diverse tasks like environmental monitoring. Notable examples of mobile robots include self-driving cars and robotic vacuum cleaners. In a broader context, mobile robots play crucial roles in aquatic, terrestrial, and aerial environments, contributing to exploration, research, and surveillance.

In the aerial realm, robots can be further categorized as fixed-wing and rotary-wing aircraft, encompassing heavier-than-air planes and lighter-than-air balloons.

DOI: 10.1201/9781003522102-10

Terrestrial robots exhibit diverse forms, utilizing legs, wheels, or tracks for mobility (Ben-Ari and Mondada 2018). Asimov (2014) played a pivotal role in conceptualizing and popularizing the field of robotics through his science fiction works and short stories.

The robot has different uses in a variety of industries:

- **Military robots:** These include drones, navigators, researchers, warriors, and any other robotics applications that might be used in espionage missions and combat zones (Springer 2013).
- **Industrial robots:** They comprise of arms, grippers, and all warehouse robotics utilized for industrial process automation. They speed up production, ensure operational efficiency, and save money (Nof 2008).
- **Cobots:** Recently introduced to the market, cobots, short for collaborative robots, serve as automation coordinators in a home or public spaces such as shops (Gerhart 1999). They are equipped with advanced programming and sensors that enable them to work alongside humans on various tasks and avoid collisions.
- **Medical robots:** These are programmable equipment used in hospitals to assist with surgery since they enable precise and minimally invasive treatments (Sood and Leichtle 2013; Lendvay 2008).
- **Environmental and alternatively powered robots:** These robots utilize renewable energy sources like solar, wind, and wave energy to provide power indefinitely and to enable use in off-grid regions (Hanson 2016).
- **Healthcare robotics:** Robotics utilized for patient monitoring and evaluation, delivering medical supplies, helping medical professionals in special ways, especially collaborative robots, and robotics for prevention (Lund 2015; Lund and Jessen 2014; Okamura et al. 2010).
- **Body-machine interfaces:** These assist amputees in receiving sensory feedback that transforms digital readings into feelings and feed-forward controls that detect their desire to move (Coates 2008).
- **Humanoids:** These offer robots expressions and responses similar to people's and combine with artificial intelligence and machine learning technology (Kajita et al. 2014).
- **Space robots:** These are robotic equipment utilized in space missions, which are quite durable and skilled in exploration and gathering material data (Launius 2008).
- **Welding robots:** This is the application of robots in the welding process. It enables automation in the welding industry (Goldiez 2021).
- **Assembling robots:** Assembling robots are robots capable of constructing different products. They also have a range of capabilities to position, mate, fit, and assemble components or parts for a product (Steven Douglas Corp 2022).

Furthermore, robots are made up of two major components: actuators and sensors.

8.1.1 Actuators

A robot achieves movement through actuators, often referred to as effectors. Actuators enable robots to manipulate objects and navigate their surroundings. The term "locomotion" encompasses the various ways a robot can move within its environment. Robots exhibit locomotion through actions such as walking, leaping, trotting, hopping, and various gestures, enabling them to traverse from one point to another. Robots contribute to human activities by undertaking monotonous and physically demanding tasks without experiencing fatigue. They can detect, manipulate, alter physical attributes, and even dismantle objects (Tirgul and Naik 2016).

In addition to electric actuators such as stepper motors, servo motors, and DC motors, other types of actuators include hydraulic, pneumatic, and artificial muscle systems (Grupen 2015). Some actuators are designed for specific tasks, such as locking and unlocking doors, activating and deactivating lights and electrical devices, issuing alerts and messages to users for potential hazards, and regulating home temperatures. Collectively, these actuators form an integral part of a robotic system (Sethi and Sarangi 2017).

8.1.2 Robot Sensing Ability: Sensors

In using sensors, robots employ either proprioceptive or exteroceptive sensing methods. Proprioception allows robots to detect their physical features, including joint angles, motor speed, battery voltage, and battery status. Exteroception, on the other hand, refers to a robot's ability to perceive its environment, such as determining distances to objects (Everett 1995). Consequently, sensors can be categorized into two groups: active and passive. Active sensors modify their environment by emitting energy, while passive sensors absorb energy from the environment to observe (GISGeograph 2015).

Furthermore, various devices can be utilized to operate robots, with Android smartphones currently being the most widely used technology (Yeole et al. 2015). Many web applications leverage the built-in hardware of modern mobile phones, such as Bluetooth, Wi-Fi, and ZigBee technologies, for controlling other devices. Bluetooth technology, which has evolved alongside modern technology and Android smartphones, facilitates wireless data transmission at close range through radio waves. The authors developed a robot controlled by an Android phone app. Through Bluetooth, the Android phone provides instructions on the robot's movements, senses and transmits information about its location and proximity to obstacles, and controls motor speed. An Android smartphone equipped with an embedded accelerometer and Bluetooth module can employ motion technologies to capture gestures and manage the dynamics of a robot, offering ease, perception, and controllability (Sharma et al. 2014).

8.2 RISK ASSOCIATED WITH ROBOTICS

Robotic hardware designs are becoming more complicated as the variety and quantity of onboard sensors rise, and more processing power is offered in ever-smaller packages onboard robots. However, these improvements in hardware do not necessarily translate into better software for commanding sophisticated robots (Sofge et al. 2003). Due to the rapid development of technology, mankind has recently attained outstanding achievement and growth. Robotics is one of these technologies; it has been around for almost a century and is still developing (David et al. 2022). Its evolution ranges from basic remote-controlled systems to humanoid robots. Security risks have been created for various purposes like defense, home, medical, or space. As these fields involve sensitive and intricate corporate tasks, protecting systems against their exploitation is of the utmost importance. As applications and accuracy have increased for every new system implemented, security risks have also been created (Priyadarshini 2017). But as technology advances, there are driverless cars using their own and cell phones that can accurately measure the heart rate of an individual. With such unique characteristics, there is a higher likelihood of undiscovered dangers. Smart devices can now be eavesdropped on and used to access private data. Recently, a survey revealed that over 6900 devices had been wirelessly compromised (Cortney 2015).

However, cyber security dangers have existed for robotics since its creation. Any cyber attack in the robotics field is based on the compromise of an endpoint or network communication. While a network communication-based assault invites an attacker to either spy on the network or inserts malicious code into it, an endpoint breach prevents a controller from controlling the robot. Physical access is a factor that contrasts the strength of the two assault vectors. Physical access attacks are far more practical than endpoint compromise attempts because they rely more on network communications (Tamara et al. 2015).

In the study proposed by Tamara et al. (2015), the robotics arena is threatened by many risks and vulnerabilities. The attacks faced by robots include the following:

Intention modification attacks: These attacks are carried out to influence a robot under the controller's control. While the packets are still in a transition stage, this specific attack seeks to change the message. Specifically, an attacker modifies packet headers to either change the packets' destination or alter the data on the target machines. Attacks with a denial of service exemplify an intention modification attack. The network interface of the robots is overloaded with transmission control protocol traffic during such attacks. It might lead to one of two outcomes:

i. **Robot halting:** It is a tactile cue that a denial-of-service attack has rendered the robot helpless. It might potentially cause the robot to move erratically. The robot may pause multiple times and for varying lengths of time. The speed might also change.
ii. **Delay in responding to direction commands:** When switching from low to fast speed, a robot under denial-of-service assault exhibits delay, as different navigational commands might not be immediately acted upon.

Intention manipulation attacks: Attackers who use intention manipulation reconstruct the message sent from the robots to the controller. These are also feedback messages because they respond to the controller's input. They could take the shape of readings or video clips. There is some level of trust because the controller's intention is genuine. It is challenging to carry out these attacks. However, identifying or stopping such attacks might be challenging if carried out well. An unfavorable outcome could result if the modified input is true. Robots are typically controlled through communication networks, making them highly susceptible to manipulation attacks. Programming a worm to spread over the network and alter robot parts without human participation is possible.

Hijacking attacks: It is a form of attack where the enemy seizes command of the lines of communication between the two endpoints. If the opponent assumed that the controller and the robot were the endpoints, it is possible that they would reject the controller's intentions and carry out unethical deeds. The services may be interrupted for a few hours or permanently depending on how long the robot is under temporary or permanent control during the hijacking. A hijacking attempt on a robot is demonstrated via the hacking of the transported surgical robot Raven II. Two different types of attackers may carry out the attacks mentioned here. The following is a list:

- **Network observer:** An adversary intending to listen in on or snoop on the data sent between a controller and a robot. He can gather and introduce false information into the communication network while appearing innocent to both sides.
- **Network intermediary:** An adversary positions himself between the controller and the robot, thus preventing confidential communication between the ends.

Moreover, the increase of various robotic security and cyber-security issues, threats, and vulnerabilities, in addition to their harmful effects, are shown as follows:

- **Security and system flaws:** These risks affect industrial robots' standard processing and performance and could disrupt production and industrial processes, leading to financial losses. However, they could result in system blockage, data interception, extraction, and physical damage.
- **Supply-chain disruption:** disrupting semi- or fully automated supply chain systems may lead to drastic financial losses and significant time-to-repair, risking the availability of robotic services and activities (Salamai et al. 2019).
- **Device theft:** Robotic devices are also prone to physical theft or hijacking and control; a prime example is a de-authentication process

that allows malicious users to disconnect legitimate owners and re-control them (i.e., robots and drones) (Yaacoub et al. 2022).

- **Network connectivity:** This is linked to the cloud decentralization strategy, which helps reduce denial of service attacks. However, it comes at the cost of reduced resource elasticity and targeted attack behaviors. Moreover, it also risks affecting supply chain management and disrupts the agility of supply chains (Sobb et al. 2020).
- **Track and trace problems:** This risk can affect the real-time ability to locate robotic transits and deliveries. This may lead to supply chain poisoning and reduced performance, especially with adopting 5G technology (Sobb et al. 2020).
- **Back-doors:** Ill-configured robotic applications with third-party access led to various backdoor and rootkit attacks. This would expose robotic users by targeting their privacy first and then by keeping them under constant surveillance, monitoring, and tracking, with the possibility of registering keystrokes and capturing snapshots or even videos without their knowledge (Cerrudo and Apa 2017).

8.3 RISK MANAGEMENT TECHNIQUES IN ROBOTICS TECHNOLOGY

A robot's security challenges are comparable since the robotics platform is built on the foundation of traditional computer systems. The significant components of robotic platforms are hardware and software. As general-purpose robots gain popularity, numerous apps emerge that let the robots carry out specific jobs. Consequently, it is necessary to secure the robot. While the system can be connected using high-level abstractions, privacy is another critical issue that can be addressed with specific access control mechanisms. Hence, in addressing these various risks and challenges, the following risk management processes aid in preventing risks and ensuring appropriate and timely responses to mitigate risk impact and severity.

8.3.1 General Risk Management Process

a. **Hazard and risk identification:** This is the first phase of risk management in robotics is the assessment of hazards and risk identification. This phase is conducted manually by identifying and describing all possible robot threats. There are several ways to perform this phase, such as fault tree analysis (FTA), and failure modes and effects analysis (FMEA).
b. **Risk analysis:** The risk analysis phase comprehends the nature of risks and determines the level of risk, including risk estimation alongside its severity (ISO 2010). In this phase, key entities, attributes of the entities, and the relationships among the attributes are identified, and then risk estimation is performed (Inam et al. 2019). Furthermore,

risk analysis can be either qualitative or quantitative. Qualitative risk analysis assesses the priority of identified risks using their probability of occurrence and the corresponding impact on project objectives if the risks do occur. While quantitative risk analysis is performed on risks prioritized by the qualitative risks analysis process as potentially and substantially impacting the project's competing demands, it analyzes the effect of those risk events and assigns a numerical rating to those risks (Constantin and Bogdan-Ion 2010). Furthermore, the last three phases of the risk management process are performed inside the robot's control loop, as shown in Figure 8.1.

c. **Risk evaluation:** The risk evaluation phase identifies the aggregation, grouping, and consolidation of the risk analysis results determined by the likelihood and consequences shown in the risk matrix. The risk matrix accounts for how the risks are evaluated in relation to their likelihood, whether critical or minor. More so, the risk evaluation is executed easily by plotting each risk in the risk matrix and adjusting and updating the results from the previous step. The aggregation aspect deals with the combination of both the higher and lower risks from the main consequences to make the combined risks a single risk. Moreover, the aggregation of risks might affect a single asset or even more. After the risks have been evaluated, they will further be classified as either consisting of risk or not; this aspect defines the risk evaluation grouping and also grouping the evaluated risks serve as a corrective measure for the identified risks, thereby ensuring that the risks are brought to view and treated. Consolidating the risk analysis result is the final aspect of risk evaluation.

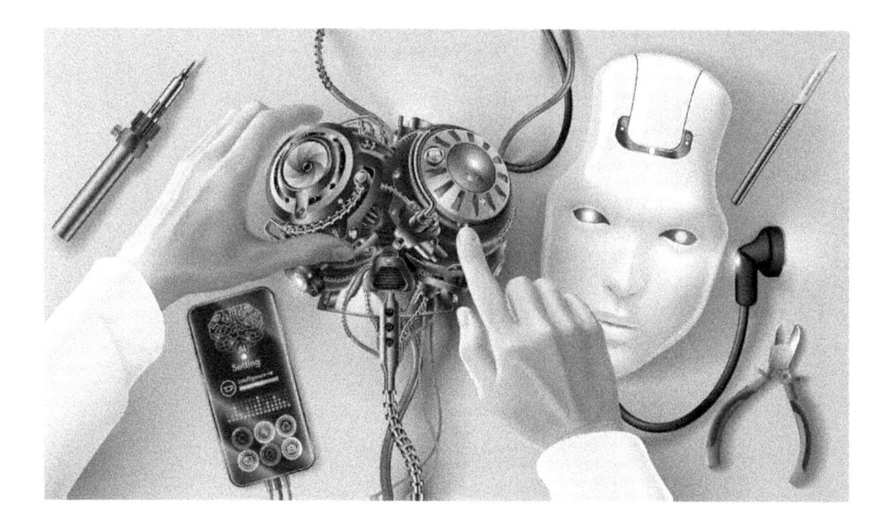

FIGURE 8.1 Risk management process, its components, and inputs/outputs (Inam et al. 2019, p. 746)

It aims to ensure the corrections to all risks are documented to enhance proper treatments. In addition, this aspect plays an important role in risk evaluation in which the treatments are identified, and essential provisions for management are applied as well as updating and correcting the risk analysis (Refsdal et al. 2015).

d. **Risk mitigation (risk reduction or treatment):** After identifying the risks, a scheme for correcting the risks is applied. This scheme contains the treatments for the risk analysis. In a nutshell, risk treatment is a process involving the implementation of modification and reduction of risks. The risk analysis could be corrected using this scheme depicted in Figure 8.2. Avoiding risky activities and careless use of damage-causing objects eradicate the risk. Also, risk mitigation reduces the threats and prevents damage; thus, the damage-causing objects are maintained and modified. In addition, the risks could be reduced by redirecting them to someone else, that is, objects in question could be insured, put on sale, and passed to investors. The exposure of the firm reduces the dominance of risks against competitors.

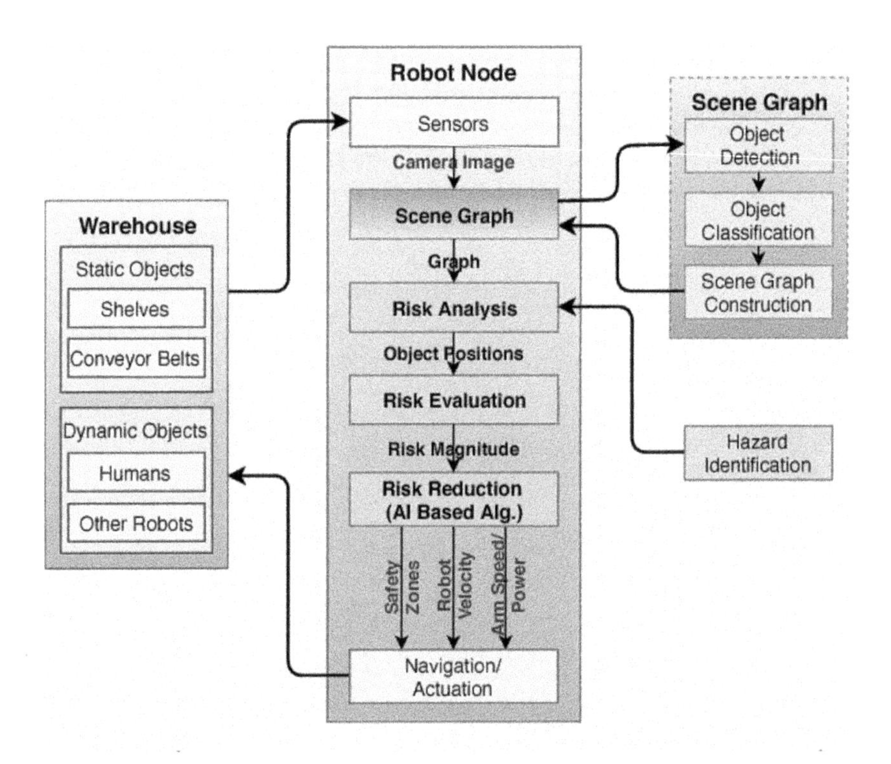

FIGURE 8.2 Selection and implementation of measures to reduce risk (Krzemien et al. 2016, p. 1046)

8.3.2 HAZard OPerability Analysis

This is a systematic and structured strategy to detect risks. This method primarily concentrates on operational risks. However, modern methods are the only ones that can limit a model's complexity. Additionally, Inam et al. (2019) assessed human-robot collaboration using this risk management technique.

8.3.3 Preliminary Hazard Analysis

This is a simple but inductive method in which hazards for a specific scenario are identified from hazard checklists of a standard (e.g., ISO12100) (ISO 2010). This entails identification of potential and actual list of risks from different sources and through the prism of different parameters associated with the technology. Thereafter, categorize the hazards attached to this risk, which provides a basis and rational for control measures.

8.3.4 The Risk Assessment Process

This technique follows three stages, as seen in Figure 8.3. In this risk assessment process, there are generic inputs that must be analyzed across the three stages. These include the enterprise environmental factors (EEFs), the organization process assets (OPAs), the project scope statement, the risk management plan, and the project management plan. The EEFs are external factors that exhibit influence on the organization process that are beyond the control of an organization or a project team, stemming from different environmental factors like political climate, legal landscape affecting the business, business associations, regulations, financial conditions in a country, market conditions and cultural landscape among others. These areas are potential areas for identifying risks that may affect the business climate of the robotic project. Hence, an environmental business scan must be conducted to identify risk areas and items in the EEFs between project initiation and the justification of the business case. However, there are schools of thought that there are internal EEFs that have led to the OPAs, which entail the intellectual, knowledge, culture, internal process, and resource repositories of an organization that affects its performance or determines the level of project success. In the context of a robotic project, there should be a compliance assessment process for a project in parri-passu of the OPA to identify risk and aid in quantitative and qualitative risk analysis.

Additionally, an input across Figure 8.3 is the project scope statement, which outlines the boundaries and rules of engagement of a project. Identifying and analyzing risks can be achieved via a two-column of activities or items within and outside the scope of a project, with appropriate justification. This ensures that project execution does not encounter bottlenecks, waste resources, and attract unnecessary processes. Hence, when a project is out of scope, there are possibilities of risks, which can then be analyzed and managed. The risk management plan is another input for risk identification and analysis, which is a document that

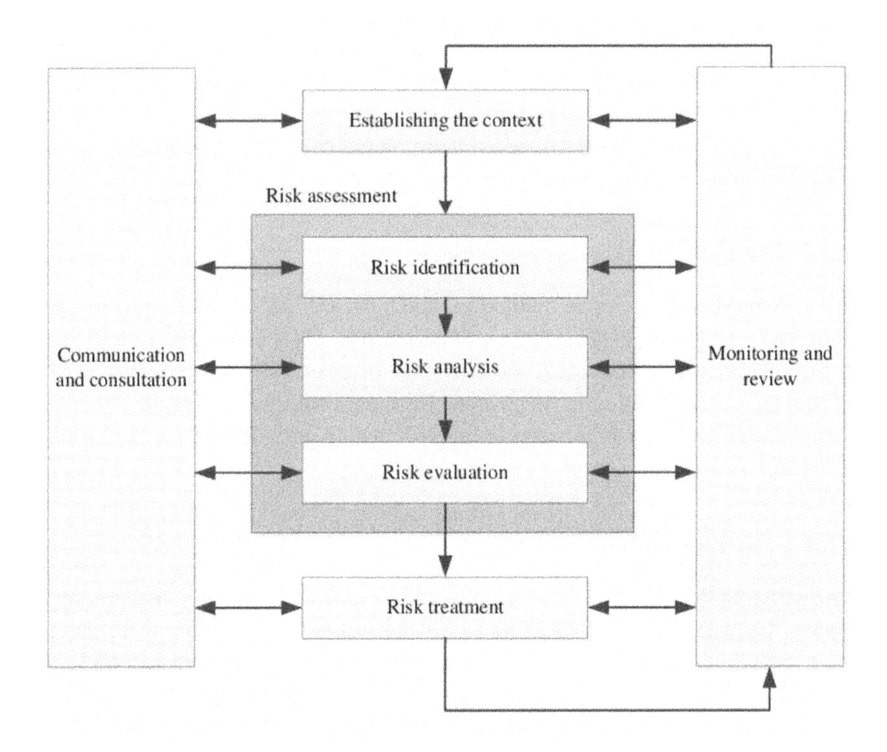

FIGURE 8.3 Risk assessment process (Ispas and Lungu 2010, p. 99)

outlines risks with their severity level and approaches to manage the risk impact. In this context of risk assessment, it is digesting and analyzing risk management plans of similar technological innovations and robotic risk management plans of previous projects. This will provide a clear insight into the landscape of robotic risks. Moreover, the project management plan in the risk assessment stages gives managerial perspectives of how a project will be executed, monitored, and controlled. This is usually dependent on the EEFs and OPAs of an organization; hence, it differs from one organization to another and from one project to another. Therefore, in identifying and analyzing risks, the project management plan is crucial in providing the context of how the project will be managed and compliance mechanisms, highlighting areas of deviation that may lead to risks.

Risk identification: This stage entails a series of processes of identifying the risks that might affect the project and understanding the various characteristics of these risks along with the inputs using the tools and techniques in Figure 8.3 to generate a risk register for the robotic project.

Qualitative risk analysis: The next stage is the qualitative risk analysis, where the risks identified are subjected to qualitative dimensions, that is

1. Risk identification	2. Qualitative risk analysis	3. Quantitative risk analysis
1. Inputs 1. Enterprise environmental factors 2. Organizational process assets 3. Project management plan 4. Risk management plan 5. Project management plan 2. Inputs 1. Documentation reviews 2. Information gathering techniques 3. Checklist analysis 4. Assumptions analysis 5. Diagraming techniques 3. Outputs 1. Risk register	1. Inputs 1. Organizational process assets 2. Project scope statement 3. Risk management plan 4. Risk register 2. tools and Techniques 1. Risk probability and impact assessment 2. Probability and impact matrix 3. Risk data quality assessment 4. Risk categorization 5. Risk urgency assessmnet 3. Outputs 1. Risk register (updates)	1. Inputs 1. Organizational process assets Project scope statment 3. Risk management plan 4. Risk register 5. Project management plan Project schedule management plan Project cost management plan 2. tools and Techniques 1. Data gathering and representation techniques 2. Quantitative risk analysis and modeling tecniques 3. Outputs 1. Risk register (updates)

FIGURE 8.4

non—numeric analysis based on facts, feelings, business psychology, observation, relatable relationships, and giving contextual meanings to the risk. Another aspect of qualitative risk analysis that should be done along the inputs is mapping out interests associated with a robotic project, which can be done by surveying and giving contextual meanings to Abraham Maslow's hierarchy of needs, McClelland's theory of needs, and Vroom expectancy theory. This will bring out behavioral risks that may lead to technological risks or corporate espionage for a robotic project. This non-numeric analysis and putting face value on robotics risks leads to an updated risk register.

Quantitative risk analysis: The third stage is analyzing the identified risks using a series of quantitative analysis and modeling software. It is also subjecting the results from the qualitative risk stage to the numeric value to understand the extent of impacts as well as the level of severity. Digital technologies of the Fourth Industrial Revolution, like the machine learning aspect of artificial intelligence and big data analytics, can be used to analyze and predict the impact and severity level of risks and prescribe appropriate solutions.

8.3.5 Robotic Risks Assessment

Robotic systems are vulnerable to several attacks. These allow attackers to gain access to systems and perform malicious tasks. However, several researchers have developed some risk assessment methods. The field of robotics has seen the emergence of numerous risk assessment and management techniques. The threat, risk, and vulnerability analysis (TVRA) methodology is the foundation for this approach (Moalla et al. 2012). This technique evaluates the effects of threats and assaults. Additionally, the concept of OCTAVE (operationally critical threat, asset, and vulnerability evaluation) was put forth by Alberts et al. (1999). This approach assesses risks based on a risk acceptance level rather than by emphasizing risk avoidance. To ensure a quantitative risk assessment of risk components, another approach known as "Methode Har-monisee d'Analyse de Risque (MEHARI)" was proposed in (Mehari 2010). It is based on assessing the maturity of the system level. The Central Communication and Telecommunication Agency (CCTA) Risk Analysis and Management Method (CRAMM) is a software application that was also introduced (Barber and Davey 1992), which helps discover and analyze threats and their impacts on information systems.

8.4 RISK MANAGEMENT SKILLS FOR ROBOTICS

Priyadarshini's (2017) study focused on robotics risk management skills. Since security lapses know no bounds and the field of robotics is rife with vulnerabilities, it is imperative to stop such attacks before they harm an individual or an organization. Several techniques have been proposed to secure a system to detect attacks and reduce threats. In today's growing technological world, the skills required to manage robotics risks are categorized into soft, hard, or technical skills. The soft skills include and are not limited to the following:

Systems thinking: To become a good roboticist, one must be skilled at systems analysis and evaluation. Understanding how robots function requires systems thinking abilities (Owen-Hill 2020).

Active learning: It is not possible to learn everything there is to know about robotics during a robotics course. Therefore, being able to learn actively means that such an individual can swiftly pick up new information as necessary (Owen-Hill 2020).

Mathematics: Math is one of the fundamental soft skills needed in robotics. However, since algebra, calculus, and geometry are the basic concepts on which robotics is based, you could struggle with some ideas in robotics if you don't have a prior understanding of these subjects (Owen-Hill 2020).

Science or other applied mathematics: Robotics is a practical application; hence, understanding applied mathematics is crucial. Applied mathematics is significant because the real world and precise mathematics are not the same. Consequently, a roboticist must determine when a calculation's outcome is accurate enough for use (Owen-Hill 2020).

Judgment and decision-making: Analytical and critical thinking abilities make up this competence. A risk manager in a robotic technology project's analytical thinking abilities will enable such an individual to examine a problem from various angles. In contrast, critical thinking abilities will help the individual to balance a solution's strengths and weaknesses (Owen-Hill 2020).

Good communication: As a roboticist, effective communication is essential. This is because of the frequency of explanations to those who aren't experts. The capacity to communicate clearly, both orally and in writing, is required to manage robotic risk (Owen-Hill 2020).

Complex problem-solving: Foreseeing difficulties, or the capacity to identify them in advance and address them as soon as they do, is a necessary skill for problem-solving (Owen-Hill 2020).

Persistence: Persistence is a crucial quality in robotics because of its complexity. Therefore, the capacity to persevere through discovering a complex problem's solution and attempting to communicate it to others (Owen-Hill 2020).

On the other hand, the hard or technical skills required for robotics include and are not limited to the following:

Micro-robotics: Today's globe has seen a significant rise in micro-robotics, as there is a lot of interest in the topic of micro-robotics from a variety of sources, including engineers who specialize in the field, engineers from other fields of engineering, scientists, hobbyists, and even members of the public (D'Souza et al. 2016).

Digital electronics and microprocessors: An in-depth understanding of designing, constructing, and testing digital logic circuits, systems controlled by microprocessors, and embedded systems is provided through the course digital electronics and microprocessor (Lawrence Technological University 2022).

Technology design: This expertise guarantees the expert creation of flawless systems. Additionally, it concerns the capacity for "repairing" or seeing flaws (Owen-Hill 2020).

The programming mind-set: This is distinct from knowledge of programming languages. This type of soft talent ensures that one's mind is prepared to learn any programming language when necessary (Owen-Hill 2020).

Machine learning: Building machines that can predict data patterns requires this ability. However, it can encourage the automation of robotics when combined with artificial intelligence.

8.5 SUMMARY

The chapter introduced robotics technology and its areas of application. It also identified the various risks attached to the utilization of the technology. The risk management techniques for robotics were discussed. These techniques include hazard or risk identification, analysis, evaluation, and treatment. The other techniques discussed were HaZard OPerability analysis, preliminary hazard analysis, the risk assessment process (subdivided into risk identification, qualitative risk analysis, quantitative risk analysis), and robotic risks assessment (quantitative risk assessment methods). Under the robotic risks assessment methods, the TVRA methodology, OCTAVE, Methode Har-monisee d'Analyse de Risque (MEHARI), and the

CRAMM were discussed. Finally, the risk management skills for robotics were categorized into soft skills and technical skills. The discussed soft and technical skills include and are not limited to the following: systems thinking, active learning, mathematics, science or other applied mathematics, judgment and decision-making, good communication, complex problem solving, and persistence skills. The technical skills discussed for managing risks in robotics include micro-robotics, digital electronics, microprocessors, artificial intelligence, technology design, programming mind-set, and machine learning.

REFERENCES

Alberts, C., Behrens, S., Pethia, R., and Wilson, W. (1999). Operationally Critical Threat, Asset, and Vulnerability Evaluation (OCTAVE) Framework, Version 1.0. https://resources.sei.cmu.edu/library/asset-view.cfm?assetid=13473.

Asimov, O. (2014). An Overview on the Concept and Laws of Robotics. *International Journal of Computer Applications (IJCA)* 13(2), 112–119.

Barber, B., and Davey, J. (1992). The Use of the CCTA Risk Analysis and Management Methodology Cramm in Health Information Systems. *Medinfo* 92, 1589–1593.

Ben-Ari, M., and Mondada, F. (2018). Elements of Robotics. https://doi.org/10.1007/978-3-319-62533-1_1, 1–20. Accessed 20 September 2018.

Bonaci, T., Herron, J., Yusuf, T., Yan, J., Kohno, T., and Chizeck, H.J. (2015). To Make a Robot Secure: An Experimental Analysis of Cyber Security Threats Against Teleoperated Surgical Robots. *ACM Transaction on Cyber-Physical Systems* 2.

Cerrudo, C., and Apa, L. (2017). Hacking Robots Before Skynet: Technical Appendix. White Paper, Technical Report. IOActive Security Services. USA. https://ioactive.com/pdfs/Hacking-Robots-Before-Skynet-Technical-Appendix.pdf.

Coates, T.D. (2008). *Neural Interfacing, Forging the Human-Machine Connection*. Morgan & Clayton Publishers, John D. Enderle Series Editor.

Constantin, I., and Bogdan-Ion, L. (2010). Project Risk Assessment Regarding Industrial Robots Implementation in Production Systems. *University "Politehnica" of Bucharest Scientific Bulletin, Series D: Mechanical Engineering* 72(2), 98–105.

Cortney, L.B. (2015). *Cybersecurity Challenges for Manned and Unmanned Systems*. West Virginia, United States: American Military University, Homeland Security.

D'Souza, R.D., Sharma, S., Pereira, A.J., and Al-Hashimi, A. (2016). Microrobotics: Trends and Technologies. *American Journal of Engineering Research (AJER)* 5(5), 32–39.

David, L.O., Nwulu, N.I., Aigbavboa, C.O., and Adepoju, O.O. (2022). Integrating Fourth Industrial Revolution (4IR) Technologies into the Water, Energy & Food Nexus for Sustainable Security: A Bibliometric Analysis. *Journal of Cleaner Production* 363. https://doi.org/10.1016/j.jclepro.2022.132522.

ENISA. (2022). *Compendium of Risk Management Frameworks with Potential Interoperability*. Athens, Greece: European Union Agency for Cybersecurity (ENISA). https://www.enisa.europa.eu/publications/compendium-of-risk-management-frameworks/@@download/fullReport

Everett, H.R. (1995). *Sensor for Mobile Robot*. Boca Raton: A K Peters/CRC Press.

Gerhart, J. (1999). *Home Automation and Wiring*. McGraw Hill Professional.

GISGeograph. (2015). Passive vs Active Sensors in Remote Sensing. https://gisgeography.com/passive-active-sensors-remote-sensing/. Accessed 2 March 2020.

Goldiez, R. (2021). Welding Robots: Types, Advantages, and Limitations. https://blog.hirebotics.com/guide-to-welding-robots. Accessed 19 October 2022.

Grupen, R. (2015). Actuators. http://www-robotics.cs.umass.edu/~grupen/603/slides/ACTUATORS.pdf. Accessed 3 March 2020.

Hanson, R. (2016). *The Age of Em: Work, Love and Life When Robots Rule the Earth*. Oxford: Oxford University Press.

Inam, R., Raizer, K., Hata, A., and Souza, R. (2019). Risk Assessment for Human-Robot Collaboration in an Automated Warehouse Scenerio. *Conference: 2018 IEEE 23rd International Conference on Emerging Technologies and Factory Automation (ETFA)*, Italy. http://dx.doi.org/10.1109/ETFA.2018.8502466.

ISO. (2010). *ISO 12100: 2010 Safety of Machinery—General Principles for Design—Risk Assessment and Risk Reduction*. Geneva: International Organization for Standardization.

Ispas, C., and Lungu, B. (2010). Project Risk Assessment Regarding Industrial Robots Implementation in Production Systems. *University "Politehnica" of Bucharest Scientific Bulletin, Series D: Mechanical Engineering* 72(2), 97–106. www.researchgate.net/publication/265231093_Project_risk_assessment_regarding_industrial_robots_implementation_in_production_systems.

Kajita, S., Hirukawa, H., Harada, K., and Yokoi, K. (2014). *Introduction to Humanoid Robotics*. Cham: Springer.

Krzemien, A., Sanchez, A.S., Fernandez, P.R., Zimmermann, K., and Gonzalez Coto, F. (2016). Towards Sustainability In Underground Coal Mine Closure Contexts: A Methodology Proposal for Environmental Risk Management. *Journal of Cleaner Production* 139, 1044–1056. http://dx.doi.org/10.1016/j.jclepro.2016.08.149.

Launius, R.D., and McCurdy, H.E. (2008). *Robots in Space: Technology, Evolution, and Interplanetary Travel*. Baltimore: The Johns Hopkins University Press.

Lawrence Technological University. (2022). Digital Electronics and Microprocessor. www.ltu.edu/engineering/projects/0d324cad-6971-4fbb-9eee-d7db27f1c182/electrical-computer-engineering/digital-electronics-and-microprocessors. Accessed 25 October 2022.

Lendvay, T. (2008). Robotic Surgery Simulation: An Unintuitive Reflection. *Medical Robotics Magazine*. http://medicalrobotics.blogspot.com/2008/10/robotic-surgery-simulation-unintuitive.html.

Lund, H.H. (2015). Play for the Elderly—Effect Studies of Playful Technology. In Human Aspects of IT for the Aged Population. *Design for Everyday Life* 9194, 500–511.

Lund, H.H., and Jessen, J.D. (2014). Effects of Short-Term Training of Community-Dwelling Elderly With Modular Interactive Tiles. *Games for Health: Research, Development and Clinical Applications* 3(5), 277–283.

Mehari. (2010). *Methode Harmonisee d'Analyse de Risques*. Paris, France: Club de la Sécurité de l'Information Français.

Moalla, R., Labiod, H., Lonc, B., and Simoni, N. (2012). Risk Analysis Study of Its Communication Architecture In. *2012 Third International Conference on the Network of the Future (NOF)*, pp. 1–5.

Nof, S.Y. (2008). *Springer Handbook of Automation*. Springer Verlag. https://nivelco.com.ua/documents/technical%20publications%20docs/Nof%20S.Y.%20Springer%20Handbook%20of%20Automation,%20Springer,%202009.pdf.

Okamura, A.M., Verner, L.N., Reiley, C.E., and Mahvash, M. (2010). Haptics for Robot-Assisted Minimally Invasive Surgery. In: Kaneko, M., Nakamura, Y. (eds.) *Robotics Research. Springer Tracts in Advanced Robotics*. Berlin, Heidelberg: Springer, Vol. 66. https://doi.org/10.1007/978-3-642-14743-2_30.

Owen-Hill, A. (2020). 10 Essential Skills That All Good Roboticists Should Have. https://blog.robotiq.com/10-essential-skills-that-all-good-roboticists-have. Accessed 25 October 2022.

Priyadarshini, I. (2017). Cyber Security Risks in Robotics. In: Kumar, R., Pattnaik, P., and Pandey, P. (eds.) *Detecting and Mitigating Robotic Cyber Security Risks* (pp. 333–348). Pennsylvania, PA: IGI Global. https://doi.org/10.4018/978-1-5225-2154-9.ch022

Refsdal, A., Solhaug, B., and Stølen, K. (2015). Risk Evaluation. In: *Cyber-Risk Management. SpringerBriefs in Computer Science*. Cham: Springer. https://doi.org/10.1007/978-3-319-23570-7_9

Salamai, A., Hussain, O.K., Saberi, M., Chang, E., and Hussain, F.K. (2019). Highlighting the Importance of Considering the Impacts of Both External and Internal Risk Factors on Operational Parameters to Improve Supply Chain Risk Management. *IEEE Access* 7, 49297–49315.

Sethi, P., and Sarangi, S.R. (2017). Internet of Things: Architectures, Protocols, and Applications. *Journal of Electrical and Computer Engineering* 25.

Sha, K., Yang, T.A., Wei, W., and Davari, S. (2020). A Survey of Edge Computing-Based Designs for IoT Security. *Digital Communications Network* 6(2), 195–202.

Sharakhatreh, F. (2011). *The Basics of Robotics. Lahti University of Applied Sciences Machine- and Production Technology*. Finland: Lahti University of Applied Sciences, pp. 1–122.

Sharma, A., Verma, R., Gupta, S., and Bhatia, S.K. (2014). Android Phone Controlled Robot Using Bluetooth. *International Journal of Electronic and Electrical Engineering* 7(5), 443–448. www.ripublication.com/irph/ijeee_spl/ijeeev7n5_02.pdf.

Sobb, T., Turnbull, B., and Moustafa, N. (2020). Supply Chain 4.0: A Survey of Cyber Security Challenges, Solutions and Future Directions. *Electronics* 9, 1864. https://doi.org/10.3390/electronics9111864.

Sofge, D.A., Potter, M.A., Bugajska, M.D., and Schultz, A.C. (2003). Challenges and Opportunities of Evolutionary Robotics. *Proceedings of the Second International Conference on Computational Intelligence, Robotics and Autonomous Systems*, pp. 1–6.

Sood, M., and Leichtle, S.W. (2013). *Essentials of Robotic Surgery*. Ann Arbor, MI: Spry Publishing LLC.

Springer, P.J. (2013). *Military Robots and Drones: A Reference Handbook*. ABC-CLIO Editor, Springer.

Steven Douglas Corp. (2022). Assembly Robots 101: The What, Where, and How of Assembly Robots in Manufacturing. https://sdcautomation.com/assembly-robots-101-the-what-where-and-how-of-assembly-robots-in-manufacturing/#:~:text=Simply%20put%2C%20robotic%20assembly%20involves,or%20parts%20for%20a%20product. Accessed 19 October 2022.

Tirgul, C.S., and Naik, M.R. (2016). Artificial Intelligence and Robotics. *International Journal of Advanced Research in Computer Engineering & Technology (IJARCET)* 5(6), 1787.

Yaacoub, J., Noura, H.N., Salman, O., and Chehab, A. (2022). Robotics Cyber Security: Vulnerabilities, Attacks, Countermeasures, and Recommendations. *International Journal of Information Security* 21, 115–158. https://doi.org/10.1007/s10207-021-00545-8.

Yeole, A.R., Bramhankar, S.M., Wani, M.D., and Mahajan, M.P. (2015). Smart Phone Controlled Robot Using ATMEGA328 Microcontroller. *International Journal of Innovative Research in Computer and Communication Engineering* 3(1), 352–356. http://doi.org/10.15680/ijircce.2015.0301020.

9 Augmented/ Virtual Reality

9.1 INTRODUCTION TO AND COMPONENTS OF AUGMENTED/VIRTUAL REALITY

Augmented reality (AR) and virtual reality (VR) are dynamic technologies encompassing hardware and software components, widely adopted across diverse contexts. These technologies find application in numerous areas, enhancing user experiences in learning, practicing, simulating, playing games, and performing various tasks (Mendes et al. 2019; Adepoju et al. 2022). The underlying rationale for AR/VR applications varies, but their common thread is providing users access to information beyond their natural senses. Recognizing the potential of AR/VR to address a myriad of challenges, renowned companies like Google, IBM, Sony, and HP and numerous universities have made substantial investments in their development. Virtually all academic fields, including physics, chemistry, management, biology, mathematics, history, astronomy, and medicine, have reaped the benefits of AR/VR technologies (Rebbani et al. 2021).

Virtual reality can be described as a virtual object in a virtual environment or, more specifically, a simulation or artificial recreation of a real-life environment or situation created by a computer and intended to immerse the user by giving him the impression of directly experiencing the simulated reality, primarily through stimulation of his vision and hearing (Kumari and Polke 2019; David et al. 2022). The "virtual environment" (VE) is an excellent VR system that allows users to walk around and touch objects physically as if they were real. This environment can be navigated and possibly interacted with, simulating one or more of the user's five senses while the user is fully immersed in the real world (Rebbani et al. 2021).

VR can further be explained as a computer-generated environment that allows for communication as though it were a real environment. A good VR system would enable users to physically move about and interact with objects as if they were real (Rebbani et al. 2021). According to Bekele et al. (2018), VR can be divided into two categories: non-immersive and immersive VR; while the former is a computer simulation of the real world, the latter adds dimensions of immersion, interactivity, and user involvement, completely removing the user from his environment in the simulated reality and replacing it with a head-mounted device.

The term "augmented reality" refers to a variety of technologies that let us experience our surroundings through the lens of a digital overlay. Computer-generated perceptual information that combines the real world with a virtual one enhances the objects you perceive in the real world and enables real-time interaction and 3D registration of virtual and real objects (Pisanu and Leufer 2020). AR is a technology

DOI: 10.1201/9781003522102-11

that improves the sensory perception of reality through superimposing computer-generated virtual material over actual structure (Hugues et al. 2011; Tang et al. 2013). AR is a technology that combines VR with reality (Chen et al. 2019). It refers to a perspective of the actual physical world in which artificial input has supplemented specific features. These inputs could include audio, video, graphics, GPS overlays, and more. The AR field is still in its early stages of development. Currently, AR systems are being successfully used in some industries, including entertainment, education, rehabilitation, and the military, to mention a few. AR fuses reality and VR. The rapid advancement of AR technology has captured the public's intense interest in recent years. AR offers several advantages, including increased realism and physical interaction with virtual items. Due to mobile phones' increasing capability and popularity, AR technologies have recently acquired popularity (Yunqiang et al. 2019). The technological tools it employs include multimedia, 3D modeling, real-time tracking and registration, intelligent interaction, sensing, and more. Its fundamental tenet is applying computer-generated virtual information—such as text, photographs, 3D models, music, and video to the real-world following simulation. The two types of knowledge work together to improve the real world, complementing one another (Hu Tianyu et al. 2017).

The fundamental role of AR is to establish connections between the real environment and data produced by a device or electronic data, directly or indirectly in response to user engagement with the device. This situation provides an interface for the user of a technologically enhanced physical environment. By adding additional digital data layers, AR strives to digitally expand and integrate the user's world or the physical environment in real time. This integration can be used with a number of display technologies that can combine or overlay information (numbers, letters, symbols, music, video, and graphics) with the user's field of view of the physical world (Arena et al. 2022). AR enhances the physical world by incorporating sensory, auditory, and visual stimulation. In other words, AR fuses digital data with the user's surroundings in real time.

In contrast to VR, which produces a wholly manufactured environment, AR adds virtual information to your current natural environment. The fundamental advantage of using AR is that consumers encounter a new and improved natural environment, where the data from the virtual world is used as a tool to aid in real life. Applications for AR deliver current and highly applicable information to the purpose you are utilizing. By AR, a user's vision of the real world is improved or supplemented with additional information produced by a computer, such as computer graphics, text, sound, and other modalities. It aims to combine computer models or computer-generated data with real-world behavior surroundings collected by a mobile video camera (Albasiouny et al. 2011). Users can, therefore, naturally engage with a blended virtual and real environment using AR technology. Three qualities are essential to an AR interface: it combines real and virtual things in a real environment, acts in real-time and interactively, and stores real and virtual objects (align) with one another.

VR seeks to improve our presence in and interaction with a computer-generated environment without allowing us to interact with or see the real world, while AR

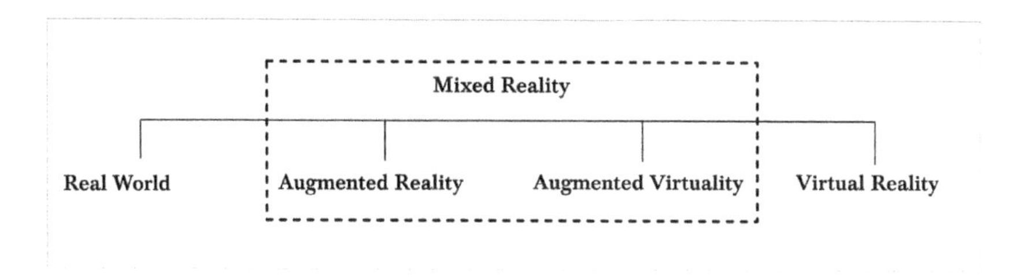

FIGURE 9.1 Augmented and virtual reality (Sünger and Çankaya 2019, p. 118).

seeks to improve our perception and understanding of the real world by superimposing virtual information on our vision of the real world (Rebbani et al. 2021). The usage of computer-generated virtual data is similar to both augmented and VR. Albasiouny et al. (2011) stated that the difference between AR and VR is that it does not attempt to conceal the user from the real world around them. Rather, it is meant to contribute more visual information to the perception of reality. To put it another way, we can say that whereas AR partially immerses the user in the virtual world, VR fully immerses the user, preserving the user's sense of presence in the real world, as depicted in Figure 9.1. Compared to VR, AR has an advantage because the user simply needs to design the virtual objects that will be incorporated into the existing world to serve as the background for the augmented entities. This can greatly minimize the time and work needed to create, render, and update the scene's computer-generated graphics.

According to Dini and Mura (2015), the following are some AR's advantages:

- Prompt interpretation of information: the user may quickly understand the messages presented.
- Immersive system: information is seamlessly incorporated into the real environment.
- Quicker procedures: The operator does not divert attention from the real world while reviewing procedural instructions.
- Paperless ability to deliver a vast amount of knowledge.
- Possibility of combining the system with other devices.

Moreover, the AR system requires specific hardware and software elements to work together, which forms its components (Kipper and Rampolla 2012).

9.1.1 Hardware

The following are the fundamental hardware elements that an AR system needs:

- **A computer or a mobile device:** In addition to creating virtual content and controlling all of the devices, it must combine the virtual content with the observer's position relative to the scene using data from the tracking system.

- **Monitor or a display screen:** Depending on the location relative to the user and the object being observed, three main types exist: (i) hand-held displays (HHD), such as a tablet or mobile phone; (ii) head-mounted displays (HMD), worn on the user's head; and (iii) spatial displays (SD).
- **Camera:** The camera is an AR system's most crucial sensor. Cameras digitally capture information about a real-world as pictures or videos, and the AR system software processes this information. An essential component of AR systems is image recognition.
- **Trackers and sensors (GPS, compass, accelerometer, etc.):** A tracking system is required to measure and record the user's position and orientation in space to align the virtual and actual images properly. Sensors are the tools used by the AR system to gather information about the real environment. GPS, a compass, an accelerometer, a camera, object detectors, and other sensors can all be used in AR systems. These sensors transmit the information they gather to the software of the AR system, where it is processed.
- **Microprocessors:** Sensor data is processed by microprocessors, which produce various textual and visual outputs and send them to displays for users to view. Computers, tablets, and mobile phones all have microprocessors. Microprocessor output must be synchronized with sensor data for them to work together. Therefore, for a flawless AR experience, microprocessor design should adhere to the specifications for AR (Craig 2013). The activities can be carried out by AR systems using a central processing unit (CPU) and a graphics processing unit (GPU) as a microprocessor.
- Interaction tools like touchpads, pointers, or wireless gadgets can be utilized as extra input devices.

9.1.2 Software

According to Kipper and Rampolla (2012), an AR system needs web services, a content server, and an application or a program running locally as its major software components. This software system enables integration with other digital technologies, components, and other systems in conveying necessary information before visualization.

9.1.3 Application of Augmented Reality

Education: According to the multimedia cognitive paradigm, people learn more effectively when several senses are engaged simultaneously. Due to its ability to appeal to a variety of people's senses, AR has a significant potential to improve education in this situation (Luckin and Fraser 2011). In addition, it is a fascinating technological advancement, and pupils presumably love using it. Additionally, it can be argued that AR offers hitherto unimaginable chances for teaching (Wu et al. 2013).

Health: According to the literature, most AR applications are in the medical or health fields. Researchers used a precise three-dimensional sensing system to create an AR navigation system that may be used during endoscopic surgery. By utilizing AR visualization, they could overlay acquired laparoscopic live images with segmented and generated CT-based images and displayed 3D-Ultrasound (US) images. The technology offers real-time anatomical data that is impossible to visualize without a navigation system. In numerous clinical cases, they also used their system. These aiding technologies are said to make the elimination of human mistakes possible (Sünger and Çankaya 2019).

Advertisement: One technology that attracts attention of advertisers is AR, which is intriguing and astounding. Advertisers have since started using AR to market their products. For clients to virtual shop for clothing online, the Smartis firm, for instance, created the Doll Up AR application. The software has a motion sensor and camera and can be operated with basic hand motions (Smartis 2012). Furthermore, marketing is successfully carried out using tools like Magic Mirror and Virtual Watch, which were created with Kinect and Zugara (Sünger and Çankaya 2019).

Maintaining and repairing: Repair and maintenance services are among the most often-used AR applications. AR is used to repair, maintain, assemble, develop, and so forth, everything from cars to cutting-edge spacecraft. It was claimed that AR for these processes saves time and money. Additionally, it gets rid of human errors that can lead to technological issues. To assist astronauts and engineers at the Jet Propulsion Laboratory, who work on the design and construction of spacecraft, the US National Aeronautics and Space Administration creates AR applications (Greicius 2016). Similarly, auto manufacturers rely heavily on AR software during the repair and maintenance processes. For instance, BMW supports its service engineers using an AR tool created to help with maintenance and repair tasks (Elearningsuperstars 2019).

Aesthetics and architecture: Architecture and interior design often utilize AR applications. IKEA Place, an augmented application created by a furniture firm, allows users to test new furniture in their homes before purchasing. Customers can use the IKEA Place AR program to design the interiors of their homes and compare various layouts with a few mouse clicks (Molla 2017).

9.2 RISKS ASSOCIATED WITH AUGMENTED/VIRTUAL REALITY

In the near future, AR and VR are expected to experience a meteoric surge in both enterprise acceptance and market size. Security experts are faced with a hurdle due to this increase in how to protect these devices and their data once they have been added to their network environments. A report by Accenture highlighted the growing number of companies taking advantage of VR and AR. It highlighted the risks that may be associated with the use of AR/VR. The associated risks include the following:

Privacy risk: Technologies like AR and VR seriously threaten privacy. The VR or AR headsets that people use for immersive training or video games have a number of sensors that capture and communicate wearer telemetry data, including body and eye movements. The ability of a device to follow a wearer's bodily movement utilizing a variety of sensors at a rate of about 90 movements per second is a prerequisite for functional AR/VR technology. This translates to recording almost two million body-movement recordings during the course of a 20-minute VR experience. Those who want to focus advertising on a particular demographic group might benefit greatly from this enormous amount of information. Customers' potential product interests can be inferred from data on the objects that users look at or interact with during the VR experience and the interaction length (Ballenson 2018). An examination of the privacy policies of certain AR/VR manufacturers reveals that aggregated data, encompassing the wearer's financial status, transaction history, and movement data, is shared with the manufacturer's business affiliates without additional consent from the wearer (Morrow 2019). The availability of telemetry data to bad actors that utilize the data to track enterprise facilities presents another threat to data privacy. This information can consist of the user's physical movements that the headset records, camera video feeds, and audio that AR/VR devices are either actively or passively gathering.

Risk of physical damage: Due to the potential legal repercussions of physical and mental injury or failure to act appropriately, physical danger can vary significantly with even brief AR/VR technologies and is possibly more significant than security. When traversing a virtual world, the wearer's eyes tell them one thing is happening, but their bodies are feeling something different or with less intensity (not moving at all). A seizure that results in serious physical harm can result from this sensory conflict, which can also make the wearer feel sick or dizzy, for example, wearer falls and receives a head injury (Lewis 2018). Additional physical risks associated with AR/VR include immersion distraction in which users lose all awareness of their surroundings as they become immersed in the virtual experience, for example, that theft or vandalism is occurring (LaMotte 2017). Loss of spatial awareness is another physical risk associated with using AR/VR. For instance, users might be unaware that they are about to run into a wall, a sharp object, or fall off a ledge, perhaps injuring themselves. Also, the risk of unwanted/unrecognized physical trauma depends on the virtual content that users are immersing themselves in; users can have hearing loss, eye damage, and behavioral problems (Kohnke 2020).

Logical and data security risk: The logical security risk associated with AR/VR devices spans a wide range of topics, including who is permitted to activate and access these devices as well as the kinds of memory devices that can be inserted into the headset and subsequently used to extract or upload data to the network. To centrally and simply manage user access and permission levels related to specific AR/VR devices, businesses should purchase AR/VR systems offering federated access capabilities. Additionally, companies run the risk of having VR headsets remotely activated while they are in use or outside of regular business hours. Providing unauthorized interior access to visitors in open spaces is another logical danger for businesses. External users digitally vandalizing internal virtual spaces (e.g., with virtual graffiti) and anonymously abusing employees in other virtual environments are risky (Fineman and Lewis 2018).

Monitoring risk: Completeness, accuracy, integrity, correlation to other network events, and logging storage capacity for events created by AR/VR devices are the main concerns for monitoring risk and controls. Creating and modifying users, applications, and requests for certain representational state transfer APIs are all events that should be monitored (Ross 2018).

Unreliable content risk: AR browsers simplify the augmentation process, but third-party businesses and applications produce and deliver the material. Since AR is still a developing field and procedures for creating and transmitting verified content are still being developed, this raises the issue of dependability. A user's AR could be replaced with their own by knowledgeable hackers, deceiving users or disseminating false information. Even if the source is reputable, a number of cyber risks may render the content unreliable. Spoofing, snooping, and data manipulation are a few of them.

Social engineering risk: Due to unreliable content, AR systems can be useful for tricking consumers as part of social engineering attacks. Hackers might, for instance, create false signs or displays to mislead users into acting in ways that are advantageous to the hackers.

Malware risks: AR hackers can insert harmful content into programs through advertising. Unaware consumers may click on adverts that take them to hostage websites or AR servers that are compromised with malware and contain inaccurate visuals, weakening the security of AR.

Network credentials theft: Android-powered wearables are vulnerable to network credential theft. Hacking may be a security risk for merchants using AR and VR purchasing apps. Many clients already store their credit card information and mobile payment methods in their user profiles. Since mobile payment is such an easy process, hackers may access these and secretly drain accounts.

Risk denial of service: Denial of service is another potential security risk on AR. For example, users who depend on AR for their jobs can experience this unexpected interruption of their information flow. In cases where not having access to information could have significant repercussions, this would be especially problematic for professionals employing technology to do duties. A driver suddenly losing sight of the road because their AR windshield transforms into a blank screen may be one scenario. Another would be a surgeon suddenly losing access to crucial real-time information on their AR glasses.

Intrusion risk: The conversations between the AR browser and the AR provider, AR channel owners, and third-party servers are susceptible to network intrusion, leading to attacks by men in the middle.

Risk of ransomware: A user's AR gadget might be compromised by hackers, who would then be able to observe and record the user's actions while in the AR environment. They might later threaten to leak these recordings to the public if the user doesn't pay a ransom. This could be embarrassing or upsetting for people who don't want their gaming and other AR interactions to be made public.

9.3 RISK MANAGEMENT TECHNIQUES IN AUGMENTED/VIRTUAL REALITY

Risk is any activity, occurrence, or choice that entails uncertainty. Risks that could negatively affect the project are only those that can be managed. Risk management includes detecting and assessing risks, analyzing risks, response planning and implementation, and continuing risk monitoring.

9.3.1 Risk Identification

Risk identification is the process of recognizing risks, which includes detecting the sources of risks, pinpointing the risk events, and ascertaining the impact of the risks related to attaining project goals (Djamaluddin et al. 2020). Risks are unavoidable and necessary in virtual and AR development. Risk identification is the first and probably most crucial step in the risk management process. Risk identification for VR/AR comes from brainstorming and experiences from similar projects. During this step, experts will brainstorm all the possible risks across all systems and then prioritize them using different factors. Another risk identification approach is through a survey, whereby users are asked about their perception of an AR device and how it affects their choices, lifestyle, comforts, and attitudes. The result from the survey could be further brainstormed for possible risks.

Identifying risks will help establish a process and responsibilities for risk management, document known risks, and allow experts to prioritize the risks. Risks can be prioritized into new and unproven technologies risks, user and functional requirements risks, application and system architecture risks, and performance or organizational risks.

9.3.2 Risk Analysis

Risk analysis is an ongoing assessment activity carried out to assess the impact of risk. Comprehensive risk analysis identifies risk factors associated with quality and safety in VR/AR. Standard risk analysis procedures in VR/AR include FTA, FMEA, formal approaches, and probabilistic safety analysis. The issue with conventional approaches like FTA and FMEA is that they must be completed manually, requiring significant time and labor for modern and complex systems. Model-driven risk analysis methods like CHESS-FLA (failure logic analysis within the CHESS project), Concerto-FLA (failure logic analysis within the Concerto project), fault propagation and transformation calculus (FPTC), failure propagation and transformation analysis (FPTA), and hierarchically performed hazard and operability studies (HiP-HOPS) have been developed for complex analysis of virtual/augmented risks.

9.3.3 Risk Response

The process of creating actions and measures to improve opportunities and lessen threats to project objectives is known as risk response planning. The main advantage of this method is that risks are dealt with according to priority, with resources and activities added to the budget, schedule, and software management plan as necessary (Rehácek 2017).

Mitigating options include the following:

- **Accept:** Acknowledge that risk will impact the VR/AR system and make an explicit decision to accept the risk without any changes to the project.
- **Avoid:** Adjust project scope, schedule, or constraints to minimize the effects of the risk.
- **Control:** Take action to minimize the impact or reduce the intensification of the risk.
- **Transfer:** Implement an organizational shift in accountability, responsibility, or authority to other stakeholders that will accept the risk.

9.3.4 Risk Monitoring

Software risk management must be integrated into VR/AR for effectiveness. This entails frequent monitoring during the design and implementation of VR and AR systems. Risk monitoring comprises reporting on VR/AR status and mentioning risk management concerns, updating risk plans in response to any significant shifts in the VR/AR schedule, examining and reordering risks, and eliminating the ones with the lowest chance.

9.4 RISK MANAGEMENT SKILLS FOR AUGMENTED/VIRTUAL REALITY

Programming skills: A basic understanding of programming languages such as Java, C#, Swift, and JavaScript and concepts like loops, control-logic, inheritance, abstraction, encapsulation, and so forth are needed to effectively identify and manage risks involved in VR/AR.

Data analysis skills: Risk managers need analytical abilities to gather and use that data to make critical decisions. They must also look for flaws and gaps in the infrastructure, systems, and other places that may have gone unnoticed by others. Risk managers must understand the effective use of enterprise risk planning systems while managing enormous amounts of data.

Decision-making skills: Risk management entails making choices that will affect risk in a foreseen and managed manner. Risk managers need to be able to rank the likelihood and severity of each risk on a scale of their choosing and make wise decisions. To make the best decision possible, people must be capable of obtaining information, weighing their options, and making a final decision.

Problem-solving skills: Risk management is a strategic business. At the upper levels, one can develop strategies and procedures for risk management throughout an entire organization.

Leadership and management skills: Effective risk managers must possess strong leadership skills to inspire and lead their teams. Leadership is crucial for navigating challenges in risk management, often requiring the disruption of established norms. To successfully manage risks, managers need to earn the respect and trust of their teams, fostering an environment where everyone feels supported. Solid management and leadership skills are essential to assist team members in handling risks within their respective domains. Additionally, adept leaders can facilitate accepting changes within the team, ensuring a smooth transition in risk management strategies.

Strategic thinking and adaptability skills: Strategic thinking involves the purposeful and cautious anticipation of risks, threats, vulnerabilities, and opportunities. Risk managers must be able to think ahead of risks and produce a distinct set of objectives, plans, and novel concepts necessary to endure and flourish in a fierce, dynamic environment. This thinking must consider market pressures, economic reality, and resource availability. Risk managers must also be able to keep up with ever-changing trends and technologies in the extended reality industries.

UI/UX designing skills: AR/VR environments often require good design and usability solutions to be fully appreciated; hence, a VR/AR risk manager must inculcate excellent designing skills. A proper understanding of UI/UX, also referred to as user interface and user experience, can also help design comfortable and efficient AR/VR headsets for users to enjoy. So, risk managers in the VR/AR industry should consider usability, accessibility, and interactions.

3D animation and modeling skills: VR/AR risk managers must be familiar with 3D software programs like Blender, Unity, and Unreal Engine because they are essential for creating environments for augmented and VR; hence, they will help identify and analyze risk.

9.5 SUMMARY

The chapter provided a detailed description and similarity of AR and VR technology. The chapter also stated the various components of the technology and its various areas of application. The chapter also outlines and explains the major 11 types of risks and risk management techniques. The chapter enumerated the multiple skills needed to manage the risks of augmented and VR technology.

REFERENCES

Adepoju, O., Aigbavboa, C., Nwulu, N., and Olaiya, M. (2022). *Reskilling Human Resources for Construction 4.0. Implications for Industry, Academia, and Government*. Springer. https://doi.org/10.1007/978-3-030-85973-2.

Albasiouny, E., Medhat, T., Sarhan, A., and Eltobely, T. (2011). Stepping into Augmented Reality. *International Journal of Networked Computing and Advanced Information Management* 1, 9–47. https://doi.org/10.4156/ijncm.vol1.issue1.2.

Arena, F., Collotta, M., Pau, G., and Termine, F. (2022). An Overview of Augmented Reality. *Computers* 11, 28. https://doi.org/10.3390/computers11020028.

Ballenson, J. (2018). Protecting Nonverbal Data Tracked in Virtual Reality. *JAMA Pediatrics* 6. https://vhil.stanford.edu/mm/2018/08/bailenson-jamap-protecting-nonverbal.pdf.

Bekele, M.K., Pierdicca, R., Frontoni, E., Malinverni, E.S., and Gain, J. (2018). A Survey of Augmented, Virtual, and Mixed Reality for Cultural Heritage. *Journal on Computing and Cultural Heritage (JOCCH)* 11(2). https://doi.org/10.1145/3145534.

Chen, Y., Wang, Q., Chen, H., Song, X., Tang, H., and Tian, M. (2019). An Overview of Augmented Reality Technology. *Journal of Physics: Conference Series* 1237, 022082. https://doi.org/10.1088/1742–6596/1237/2/022082.

Craig, A.B. (2013). *Understanding Augmented Reality: Concepts and Applications*. San Francisco, CA: Morgan Kaufmann.

David, L.O., Nwulu, N.I., Aigbavboa, C.O., and Adepoju, O.O. (2022). Integrating Fourth Industrial Revolution (4IR) Technologies into the Water, Energy & Food Nexus for Sustainable Security: A Bibliometric Analysis. *Journal of Cleaner Production* 363. https://doi.org/10.1016/j.jclepro.2022.132522.

Dini, G., and Dalle Mura, M. (2015). Application of Augmented Reality Techniques in Through-Life Engineering Services, the Fourth International Conference on Through-Life Engineering Services. *Procedia CIRP* 38, 14–23.

Djamaluddin, I., Indrayani, P., and Caronge, M.A. (2020). A GIS Analysis Approach for Flood Vulnerability and Risk Assessment Index Models at Sub-District Scale. *IOP Conference Series: Earth and Environmental Science. Volume 419. The 3rd International Conference on Civil and Environmental Engineering (ICCEE 2019),* Bali, Indonesia, August 29–30, 2019. https://iopscience.iop.org/article/10.1088/1755-1315/419/1/012019.

Elearningsuperstars. (2019). http://www.elearningsuperstars.com/project/bmw-augmented-reality-service-engineer-training/.

Fineman, B., and Lewis, N. (2018). Securing Your Reality: Addressing Security and Privacy in Virtual and Augmented Reality Applications. *EDUCAUSE Review*. https://er.educause.edu/articles/2018/5/securing-your-reality-addressing-security-and-privacy-in-virtual-and-augmented-reality-applications.

Greicius, T. (2016). 'Mixed Reality' Technology Brings Mars to Earth. www.nasa.gov/feature/jpl/mixed-reality-technology-brings-mars-to-earth.

Hugues, O., Fuchs, P., and Nannipieri, O. (2011). *New Augmented Reality Taxonomy: Technologies and Features of Augmented Environment*. Berlin: Springer.

Kipper, G., and Rampolla, J. (2012). *Augmented Reality: An Emerging Technologies Guide to AR*. Amsterdam: Elsevier.

Kohnke, A. (2020). The Risk and Rewards of Enterprise Use of Augmented Reality and Virtual Reality. *ISACA Journal* 1.

Kumari, S., and Polke, N. (2019). Implementation Issues of Augmented Reality and Virtual Reality: A Survey. In: *International Conference on Intelligent Data Communication Technologies and Internet of Things (ICICI)* (vol. 26, pp. 853–861). Cham, Switzerland: Springer International Publishing. https://doi.org/10.1007/978-3-030-03146-6_97.

LaMotte, S. (2017). The Very Real Health Dangers of Virtual Reality. *CNN Health*. www.cnn.com/2017/12/13/health/virtual-reality-vr-dangers-safety/index.html.

Lewis, C. (2018). The Negative Side Effects of Virtual Reality. *Resource*. http://resourcemagonline.com/2018/03/the-negative-side-effects-of-virtual-reality/87052/.

Luckin, R., and Fraser, D.S. (2011). Limitless or Pointless? An Evaluation of Augmented Reality Technology in the School and Home. *International Journal of Technology Enhanced Learning* 3(5), 510–524.

Mendes, M., Almeida, J., Mohamed, H., and Giot, R. (2019). Projected Augmented Reality Intelligent Model of a City Area with Path Optimization. *Algorithms* 12(7), 140.

Molla, M. (2017). IKEA'dan Artırılmış Gerçekliği Sevdirecek Uygulama [Online]. http://dergi.ituieee.com/teknolojik/ikeadan-artirilmis-gercekligi-sevdirecek-uygulama.

Morrow, S. (2019). Is the Security of Virtual Reality (and Augmented Reality) Virtual Insanity? *Infosec*. https://resources.infosecinstitute.com/virtual-reality-vr-security-concerns/#gref.

Pisanu, G., and Leufer, D. (2020). Augmented Reality & Augmented Risks: Why AR Is a Digital Right Issue. *Access Now*. https://www.accessnow.org/what-is-augmented-reality-risks/.

Rebbani, Z., Azougagh, D., Bahatti, L., and Bouattane, O. (2021). Definitions and Applications of Augmented/Virtual Reality: A Survey. *International Journal* 9, 279–285.

Rehácek, P. (2017). Risk Management Standards for Project Management. *International Journal of Advanced and Applied Sciences* 4, 1–13. https://doi.org/10.21833/IJAAS.2017.06.001.

Ross, A. (2018). Solving the Virtual Reality Storage Challenge. *Information Age*. www.information-age.com/virtual-reality-storage-challenge-123473300/.

Smartis. (2012). Türkiye'de Bir İlk: Doll Up [Online]. www.smartis.com.tr/blog/?p=1394.

Sünger, I., and Çankaya, S. (2019). Augmented Reality: Historical Development and Area of Usage. *Journal of Educational Technology and Online Learning* 2, 118–133. https://doi.org/10.31681/jetol.615499; https://files.eric.ed.gov/fulltext/EJ1308502.pdf.

Tang, J.B., Amadio, P.C., and Boyer, M.I. (2013). Current Practice of Primary Flexor Tendon Repair: A Global View. *Hand Clinics* 29, 179–189.

Tianyu, H., Quanfu, Z., and Shen Yongjie, D.H. (2017). Overview of Augmented Reality Technology. *Computer Knowledge and Technology* 34, 194–196.

Wu, H.K., Lee, S.W.Y., Chang, H.Y., and Liang, J.C. (2013). Current Status, Opportunities and Challenges of Augmented Reality in Education. *Computers & Education* 62, 41–49.

Part III

Risk Management Framework for Fourth Industrial Revolution Technologies

10 Risk Management Framework

10.1 OVERVIEW OF RISK OF FOURTH INDUSTRIAL REVOLUTION TECHNOLOGIES

The Fourth Industrial Revolution has brought a significant change in globalization. As a result, global value supply chains become more efficient, transportation costs are reduced, and new and current markets are created and expanded, which is a key factor in driving economic growth. According to a study by the World Economic Forum, "the Fourth Industrial Revolution will raise the level of global income and improve people's quality of life worldwide" (Mitrovic 2020). However, as this technology advances, people are vulnerable to a variety of security concerns, problems, and dangers, including climate change, large-scale spontaneous migration, cyber attacks, Internet fraud, and data theft, and others, according to the 2020 Global Economic Forum's Global Risk Report (Mitrovic 2020). Digitalization, a technology integration that converges the physical, biological, and digital worlds, enables the Fourth Industrial Revolution (Industry 4.0 or 4IR). These advancements result from combining a number of cutting-edge technologies used in several academic disciplines, including nanotechnology, biotechnology, cognitive science, and information and communication technologies. Artificial intelligence, the foundation of future industries' and factories' intelligent systems, makes these technologies possible (Gillieron 2019).

Additionally, the growth of disruptive technologies will cause a wave of globalization to intensify, giving rise to the idea of "Globalization 4.0." (Gillieron 2019). Globalization and technology, however, are intimately intertwined and have a significant role in creating opportunities and hazards (Ljajic et al. 2016). The advancement brought about by the emergence of industry 4.0 marks the pinnacle of globalization (WEF 2019). The process of globalization leads to human exploitation and brain drain. However, the shift in transnational production to nations with inexpensive labor and the focus on maximizing profits raised the possibility of seeing a growth in inequality and a significant gap between the rich and the poor (Skorodumova and Melikov 2020).

The Fourth Industrial Revolution, also known as Industry 4.0, has its roots in a government-funded paper prepared in Germany in 2011 and outlined a strategy to advance manufacturing's competitive advantages through the use and integration of new technology (Bittencourt et al. 2021). Then, across all industries, its development has been linked to digitization, flexibility, automation, virtualization, resource efficiency, and decentralization (Luthra et al. 2020). This technology ensures the networking of devices in enterprises for optimum output using

DOI: 10.1201/9781003522102-13

artificial intelligence, digital twins, blockchain, and cloud computing (Ghobakhloo et al. 2021; David et al. 2022; Adepoju et al. 2022).

With the ability to enhance the working environment and product quality, these technologies can assist companies in achieving tangible results in a short amount of time (Kamble et al. 2020). It will also lead to numerous businesses benefiting from the Fourth Industrial Revolution's technology in the economy, society, and environment (Blunck and Werthmann 2017). It is used in daily operations by a variety of industries, including healthcare (Magid et al. 2021), information technology (Sobb et al. 2020), and transportation (Mohebbi et al. 2020). The Fourth Industrial Revolution will also affect the labor market, inevitably leading to an increase in inequality. The gap between the income of capital and labor income widens due to the existence of automation technologies that substitute humans for machines (Goncharov 2020).

Moreover, with the advent of the Fourth Industrial Revolution, new technologies are emerging and connecting, posing hazards to manufacturers and sectors. These hazards are caused either by the nexus between new and legacy systems or by upcoming technology. Cyber attacks are one of the risks faced by the industrial industry. These attacks target industrial control, business, or safety instrumentation systems. In 2018, Make UK and AIG research found that 48% of UK manufacturers have experienced a cyber attack, with "financial impact" being the main negative effect. Intellectual property is a common factor that motivates malevolent attackers to assault the industrial sector (Copic and Leverett 2019).

10.2 LEGAL, POLITICAL, AND SOCIAL-ENVIRONMENTAL RISKS

Due to the widespread adoption of new networked social technologies, the creation of self-learning software based on artificial intelligence, and the virtualization of the human living environment, globalization presents numerous social risks in the age of the Fourth Industrial Revolution. The potential of big data and artificial intelligence developments are firmly entwined with the cultural upheavals brought about by postmodernism, which is centered on the deconstruction of a person's psyche (Skorodumova and Melikov 2020). The widespread usage of crowdsourcing technology diminishes the importance of professional ethics and a scientist or engineer's moral obligation to account for the effects of employing their innovations. The dangers of combining big data technologies with criminal activity are growing (Skorodumova and Melikov 2020). Globalization underwent a tremendous upheaval at the time of the Fourth Industrial Revolution.

Additionally, the Fourth Industrial Revolution's advancements in robotics technology, a subspecialty, have increased unemployment rates and made it possible for individuals to work in any environment, leading to servitude to technology. Furthermore, inequality has risen significantly in rich and emerging nations due to the Fourth Industrial Revolution. Emerging artificial intelligence technology can now scan human footprints and biometrics, which gather data that can help the system detect people's issues and passions and purposefully affect people to make money or super profits (Goncharov 2020).

Also, dangerous religious groups like ISIS heavily utilize the risks of digital footprint analysis for their recruitment efforts. More specifically, cybercriminals today leverage emerging networking technologies like blockchain and self-learning programs that rely on artificial intelligence to carry out their exploitation. Criminal gangs are now purchasing databases and building strong teams of programmers to modify software from open access and obtained through frontmen from top firms. These activities point to criminal organizations' utilization of big data and the creation of novel artificial intelligence-based tactics. The loss of national-cultural identity is one risk of the globalization processes brought forth by the Fourth Industrial Revolution. This technology makes it possible to erase an entity's identity, including a person. One's individuality loses its integrity. For example, a person may become an atheist and cosmopolitan today, and the next instant, may see them appear to be a virtual identity well enmeshed in the "deep web" (Kuzmenko et al. 2018). VR technology has also helped to further the practice of changing one's consciousness and entering an unknown condition that is frequently referred to as a trance. VR technology has been extensively used in Western Europe to alienate the body, including out-of-body experiences (Ehrsson 2007). Man creates the false reality that his body exists apart from him. The illusion that the body may have awareness all by itself causes the integrity of the personality to be destroyed. It is feasible to exert total control over a subject's attention while fully immersed in VR, according to experiments involving studying human behavior in that environment (Zinchenko et al. 2010).

Industry 4.0 can be viewed as a change in strategies, organizations, business models, value and supply chains, processes, goods, skills, and stakeholder interactions to examine the legal risks and challenges of the Fourth Industrial Revolution, according to (Buchi et al. 2020). These modifications allow new legal protection areas and ensure the law has both conventional and innovative answers. The Fourth Industrial Revolution may bring forth the following legal risks:

- Personal injury
- Damage to property
- Breach of contract
- Misuse of personal data
- Loss of control of machines
- Violation of control of employee's rights
- Risk of injury or damage
- Infringement of intellectual property

(Habrat 2020)

The likelihood that a specific technological pillar will support a given legal risk and the number of operational Industry 4.0 technological pillars that support this risk will determine the probability of the risk's overall occurrence in the structure of a digital enterprise. Therefore, a significant legal danger is associated with the widespread adoption of automation and digitalization. It can be decreased by lowering the possibility of danger for a specific technology. This is generally feasible but necessitates collaboration among international legislative bodies. Unpredictability in risk areas also exists, but this is the price of industrial development. As mentioned, the Fourth Industrial Revolution has the potential to bring about globalization.

Globalization will, however, also give the international economy new opportunities. Corporate crimes such as market fraud, threats to public safety, the export of illegal goods, the provision of out-of-date and illegal goods and services online, money laundering, the use of human trafficking and white slavery, environmental pollution and degradation, and industrial espionage in the global economy are on the rise as the economy grows (Williamson 1998; Fukuda-Parr 2003; Mehdi 2013; Eldridge et al. 2017; Ignjatović 2018; Elamiryan 2019). Figure 10.1 provides a thorough picture of the security issues, dangers, and threats to societal and individual security brought on by the Fourth Industrial Revolution.

In Figure 10.1, high or erroneous investment is listed as an economic risk associated with the Fourth Industrial Revolution, whereas job losses and other social concerns are included as economic risks. However, investing in the Fourth Industrial Revolution increases the financial risk of mistakenly investing in subpar or frequently immature technologies or in processes that are not beneficial from an economic standpoint. Furthermore, economic chances will be limited because of the lack of skills required for future professions. Concerns might be connected to technological issues in addition to economic and social risks. For instance,

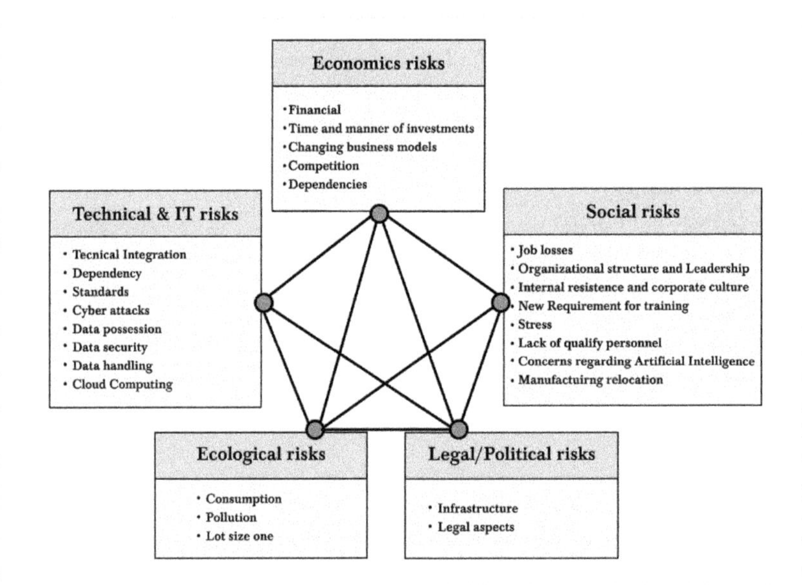

FIGURE 10.1 Main risks of the Fourth Industrial Revolution to individual and society security (Kodym et al. 2020, p. 76)

technological integration, risks associated with information and communication technologies like data security, and legal and political risks like legal ambiguity on who owns the data are still being debated. The dangers from a legal perspective include those related to data processing, protection, jurisdiction, labor legislation, intellectual property, and so forth. Additionally, there are some environmental concerns in the form of pollution and environmental degradation brought on by an increase in waste production and emissions due to the processes of the Fourth Industrial Revolution requiring more significant quantities of energy and raw materials (Mitrovic 2020).

10.3 RISK MITIGATION MEASURES

The term "risks" has no established definition as of yet. However, various authors have described it (Tang 2011; Tummala and Schoenherr 2011). Risk is "the level of exposure to uncertainties that the firm must understand and successfully manage as it implements its strategy to achieve its business objectives and produce value," according to Deloach (2000). However, the lack of agreement on what constitutes "risk" hinders practitioners' and researchers' ability to collaborate (Ho et al. 2015). Furthermore, the Fourth Industrial Revolution's technologies like cloud computing, the Internet of Things, and CPS increase risk. Understanding risk management is essential for risk mitigation. The main goals of risk management are to identify potential dangers, analyze them, and create the necessary responses (Ghadge et al. 2012; Guo 2011). Four steps make up the procedure of mitigating risks, according to Sodhi et al. (2012), which are as follows:

Identification: Risk identification is the first step in risk analysis or assessment and a crucial component of the risk management. Therefore, knowing your firm's goals is the first step in risk identification. Then, whether or not they are within your control, it should cover all potential dangers, risks, and circumstances that could jeopardize its capacity to achieve those goals. Risk identification shouldn't be a one-and-done process but rather continuous and ongoing throughout the project, program, or organization being assessed for risk because risks evolve and new risks emerge over time.

Assessment: The second procedure, "assessment," quantifies and ranks hazards. The likelihood of each risk and its related effect on the technology are confirmed during this phase. These two pieces of information are then utilized to rank the risks. The entire technological project is ongoing in this process. The risk matrix can also be used to evaluate hazards (di Lenardo 2019). However, before the execution of the technological projects, potential risks should be analyzed through a brainstorming session, evaluation of previous technological projects, and operational changes that may occur in the execution chain.

Control of hazard and risks: Risk assessment and planning for plan implementation are part of this phase. Depending on the kind and nature of the risk, it is possible to control it. But these choices could be taken into account:

i. **Avoid:** The risk is avoided by ceasing the conduct that introduced the risk. Alternatively, one might select a project that still achieves corporate goals or a better secure method or procedure.
ii. **Reduce:** The risk's probability or impact is decreased to a manageable level. According to cost and timing, it is preferable to eliminate the risk.
iii. **Share or transfer:** The risk owner no longer bears the risk. For instance, through contracting with service providers, outsourcing the operations to which the risk is linked, or purchasing insurance that addresses the risk.
iv. **Accept:** Risks with extremely low probability and consequences are frequently acceptable. Furthermore, if the expense of the treatment outweighs the benefit, a risk may be accepted. When a risk is accepted, all risk-reduction measures come to an end. Simply put, the risk is continuously monitored and assessed due to potential changes (di Lenardo 2019).

Monitoring and controls: The major goal of this step is to develop risk tracking and monitoring measures. The risk register, initiated at the beginning of a project and continually evaluated and updated, is the most frequently employed method of risk monitoring. A risk register should, at the very least, include the following details:

- Risk identification number
- Risk owner
- Description of risk
- Results of assessment (probability/impact) and assessment date
- Mitigating actions
- Date for next risk review

In a nutshell, risks need to be continuously tracked, evaluated, and managed. Risk management is carried out by continuously monitoring the known risks and finding and assessing new risks. Throughout the project life cycle, risks and the efficiency of controls and mitigation should be evaluated (Di Lenardo 2019).

10.4 RISK MANAGEMENT FRAMEWORK FOR FOURTH INDUSTRIAL REVOLUTION TECHNOLOGIES

Managing the risks of the Fourth Industrial Revolution entails the utilization of a framework which will guide each technology or combination of technologies, irrespective of their severity and level of impact. Hence, this book presents three frameworks for managing 4IR risks. Every technological project or the utilization of each of the Fourth Industrial Revolution technologies should be subjected to this analysis for technological insights on what risks to avoid, accept, mitigate, and control. The frameworks are as follows:

10.4.1 The PESTLE Analysis Framework

A PESTLE analysis as a risk management framework aims to provide insights into how a chosen technology affects the external environment of operation or execution. This is because the technology of the Fourth Industrial Revolution is seen as an "enabler," "catalyst," "optimizer," "booster," and "change agent"; hence, its effects are always felt in the environment, from production processes, products, services, human relations, and performance. Hence, performing a PESTLE analysis aids in carefully examining every detail that could jeopardize the functionality, goal, and objectives of a project or a system while utilizing the technologies of the Fourth Industrial Revolution. PESTEL is short for "political, economic, social and cultural, technological, legal, environment" (Alanzi 2018).

The PESTLE analysis is one of the most popular models for assessing the external business environment, which is very dynamic, according to Gupta (2013). It is utilized as a situational analysis tool for business evaluation reasons, with proven success, hence the adaptation as a framework for managing the technologies of the Fourth Industrial Revolution. The PESTLE framework considers an industry's external environmental elements and serves as the foundation for evaluating risks and opportunities, facilitating SWOT analysis. The use of this model is also influenced by the organization's or industry's market structure as well as by the variety of elements between industries.

The PESTLE study is a framework for looking into and analyzing an organization's external environment. However, it specifies six crucial elements that must be considered when determining the causes of potential alterations. These six elements include the following:

Political pillar: The change in government is one of the political factors. This is related to the idea that policies and projects change as the government does. In managing the risks of the technologies of the Fourth Industrial Revolution during utilization, the political climate must be understood, as political terrain impacts the utilization level. Therefore, in this political pillar of PESTLE analysis, there is a need for policy evaluation dealing with technology. This policy in terms of government regulations or bilateral convention must be analyzed when using any technology, as this will provide insights on the usage level and the hidden technological risks.

Moreover, this pillar also deals with the change of power, that is, government change, which is rooted in the ideological attitude of a government, either current or new to the use of technology in operation, and also on likely regulations and policies that might be put in place regarding technological innovation. This will guide the utilization of the technologies of the Fourth Industrial Revolution, as the mode of utilization breeds the level of risk during utilization. Another aspect of this pillar is to examine the perception of political stakeholders across the divide on the use of technology and whether they will all favor technology usage. Still, the question is to what extent will usage guide how to analyze the risks involved in a technology and also how to manage them. This analysis pillar also helps provide insurance and investment coverage when technological risks are accepted or shared. In this pillar, the following risk factors should be analyzed with consideration to their effect on 4IR technology: political stability, political stakeholders' technological ideology, government policies, corruption index, foreign policies on technology and trade, and technological restrictions policies.

Economic pillar: This pillar or factor deals with the relationship between economic activities and its effect on utilizing the Fourth Industrial Revolution technology. This has to do with the level of investment that can be deployed to technology and the rate of return from the utilization of the technology, as the rate of return from a trillion-dollar economy will be significantly higher than a rate of return from a billion-dollar economy. This investment decision always determines the extent of utilization or exploration of the technologies of the Fourth Industrial Revolution, whereby the nexus between the utilization and the risk level comes into play. Another significant feature of this pillar in managing risks is the impact of the technologies on the economy in terms of speed, adaptability, skills usage, and the level of acceptance. Certainly, the 4IR technologies will impact the economy, but the question is at what cost? What are going to be the give-and-take mechanism during utilization? The analysis of these questions will generate a series of risks and how to manage them. Another major aspect of this pillar is the economic policies in place that either support or restrain the technological usage of the 4IR technologies. Analyzing these policies will unleash the various dimensions of risks and levels of uncertainty during usage and integration, especially in the production process and sales strategy. Another dimension to consider is the impact of the 4IR technologies on a product's entire value chain and supply chain by analyzing what type of risk will emanate from the chains, leading to what decision to take to mitigate them. In analyzing this pillar with the technologies of the Fourth Industrial Revolution, the following factors should be considered: inflation rate during the purchase of technological equipment, tax rates, exchange rate, interest rate, subsidies and government incentives, unemployment rate, purchasing power, consumer price index, and consumer protection return policies.

Socio-cultural pillar: This pillar of PESTLE analysis covers the human dimension to utilizing the Fourth Industrial Revolution technologies. It encapsulates humans' effect on the technologies regarding utilization, acceptance, and skills to operate the technologies. Much research has stated that the 4IR technologies will revolutionize the way human relates, which poses the fact that there is a nexus. However, in managing the symbiotic risks between humans and technologies, the analysis must center around human potential and actual needs for the technology, technology acceptability rate, social norms, and cultural values. These factors should be evaluated, generating a list of risks in embracing the 4IR technologies. Also, in this pillar, the focus should be on the impact of 4IR technologies in social settings in terms of the relationship among people, human connection, societal connection, skill development, employment rate, population growth, health consciousness, cultural barriers, age distribution, career option and development, safety issues and modalities, lifestyles and attitudinal changes, demographics demarcation, media connections, social media effects, and peer pressure among age groups. This will lead to different types and categories of risks in implementing the Fourth Industrial Revolution, which will also generate the choice of mitigation measures in managing them.

Technological pillar: The 4IR technologies are domicile in this pillar, as previous technologies are the building blocks for the current technological wave. However, risk management has to be viewed from three perspectives. The first perspective is the incremental impact of 4IR technologies, which entails the progression by which an individual, organization, or society improves its technological status or pace due to 4IR technologies. In this phase, the various risks of incremental impact will be shown, enabling the choice of the right mitigation measures. Another perspective is the transformational impact of the 4IR technologies, whereby there is a total overhauling of existing technologies or technological methods and processes with digitalization by the 4IR technologies. It concerns the total change of existing technologies with technologies of the Fourth Industrial Revolution. This will also address utilization and change risks, which will affect the entire organization structure or societal setting. The third perspective is the vertical and horizontal integration due to the technologies of the Fourth Industrial Revolution. Organizations may adopt horizontal or vertical integration to optimize the utilization of the 4IR technologies. This will generate different aspects of risk and how to mitigate them in terms of operational efficiency, operational cost, employee knowledge, competitive advantage, societal acceptance of the technology, supply chain optimizations, value chain digitalization, profitability, market penetration strategies, production of quality goods, and flexibility of production process for mass customization. However, in addressing these three perspectives, the following technological factors should be considered: hardware sophistication, software security, technological awareness, cyber security level, research and development expenditures, employee's education and skill level, organization structure and culture, rate of training and development, technological capability level, production and competitive strategy, managerial capability, transactional capability, and operation management practices.

Legal pillar: This pillar entails the internal and external laws and regulations that guide the utilization of the 4IR technologies, which mainly entails the following: right to privacy, right to data either private or public data, patency laws, partnership deeds, information accessibility and utilization laws, copyright protection, consumer safety, international convention regulations, court conventions and practices, legislative acts and law enforcement processes. These legal factors will analyze risks and how to mitigate, control, and manage them.

Environmental pillar: The concept of sustainability, as evidenced in the United Nations Sustainable Development Goals has given more importance to the impact of technological innovation on the environment in terms of its effect on humans, ecosystems, Agricultural production, climate change, the health of man, plants and animals, biodiversity, and depletion rate of the planetary boundaries. The Fourth Industrial Revolution technologies are more about digitalizing existing technological innovation, whereby the existing innovations are environmentally disastrous. Even 4IR technologies, such as blockchain technology in cryptocurrency mining, generate huge carbon emissions. Hence, two approaches are necessary for assessing and managing the environmental risks of 4IR technologies. The first approach is the need for an environmental impact assessment of each technology during application and integration into any endeavor or business dimension. The second approach is the issue of energy utilization, whereby renewable energy utilization in powering the 4IR technologies will lead to less environmental impact.

10.4.2 The NIST Framework

In managing the various risks of the Fourth Industrial Revolution, the cyber security framework of the National Institute of Standards and Technology (NIST 2018) of the United States of America can be utilized. The framework outlines five utilization areas: identify, protect, detect, respond, and recover. It is an internal mechanism of managing risks that may want to disrupt the 4IR technologies into greater external risks.

Identify: In relation to the Fourth Industrial Revolution, the identify function of the framework entails identifying and understanding the risks to the 4IR technology in terms of data, organization and technological assets, governance structures and models within the organization, business environment, stakeholders' activities and interest and risk management strategy for each of the technology. This approach is a rigorous evaluation of the company's policies, business policies, government policies, strategic objectives of the organization, organization culture and structures, and answer to the question of what may not make the technology work as expected.

Protect: This function entails developing and implementing guidelines and strategies against risks and non-functionality of the 4IR technology. This entails the proactiveness of guiding against what was discovered in the identify function. In this function, the protection guidelines should cover access control to the technology, training on the technology, process and procedures of protecting the technology operations, the maintenance mechanism of the technology, and the authorization of usage.

Detect: This third function should cover procedures and processes in the case of a risk. At this phase, the risk is detected, whereby events and anomalies are detected, and there is continuous monitoring of the security breaches to understand what went wrong and, at what time, and who was responsible for the risks.

Respond: Once the risks have been detected, this function entails implementing strategies to contain the risk and reduce its impact. It is the remediation of the detected risk to the 4IR. This will further entail having a response plan in place, communicating to appropriate agencies, sections, or departments, mitigating the spread of the risks, and improving the architecture of the 4IR technology.

Recover: This is the function whereby the damage done in the eventuality of risks to the 4IR technology is restored, and the technological architecture is restored with more resilience to risks. In this function, there is a need for an insurance policy and backup functionality, which will address the damage of the risk.

10.5 SUMMARY

This study has presented an overview of the risks of the Fourth Industrial Revolution. Also, the legal, political, social, and environmental risks were thoroughly discussed, followed by the risk mitigation measures. Furthermore, the chapter presented two frameworks for managing the risks of the Fourth Industrial Revolution: the PESTLE analysis framework and the NIST cyber security framework.

REFERENCES

Adepoju, O., Aigbavboa, C., Nwulu, N., and Olaiya, M. (2022). *Reskilling Human Resources for Construction 4.0. Implications for Industry, Academia, and Government*. Springer. https://doi.org/10.1007/978-3-030-85973-2.

Alanzi, S. (2018). Pestle Analysis Introduction. www.researchgate.net/publication/327871826_Pestle_Analysis_Introduction.

Bittencourt, V.L., Alves, A.C., and Leão, C.P. (2021). Industry 4.0 Triggered by Lean Thinking: Insights From a Systematic Literature Review. *International Journal of Production Research* 59, 1496–1510.

Blunck, E., and Werthmann, H. (eds). (2017). *Industry 4.0-An Opportunity to Realize Sustainable Manufacturing and Its Potential for a Circular Economy*. Dubrovnik: Sveuˇcilište u Dubrovniku.

Buchi, G., Cugno, M., and Castagnoli, R. (2020). Smart Factory Performance and Industry 4.0. *Technological Forecasting and Social Change* 150. https://doi.org/10.1016/j.techfore.2019.119790.

Copic, J., and Leverett, E. (2019). *Managing Cyber Risk in the Fourth Industrial Revolution: Characterising Cyber Threats, Vulnerabilities and Potential Losses*. Cambridge: Cambridge Industrial Innovation Policy, University of Cambridge, pp. 1–22.

David, L.O., Nwulu, N.I., Aigbavboa, C.O., and Adepoju, O.O. (2022). Integrating Fourth Industrial Revolution (4IR) Technologies into the Water, Energy & Food Nexus for Sustainable Security: A Bibliometric Analysis. *Journal of Cleaner Production* 363. https://doi.org/10.1016/j.jclepro.2022.132522.

Deloach, J.W. (2000). *Enterprise-Wide Risk Management: Strategies for Linking Risk and Opportunity*. London: Financial Times/Prentice Hall.

Di Lenardo, S. (2019). Risk Management in Industry 4.0. http://wiki.doing-projects.org/index.php/Risk_management_in_industry_4.0. Accessed 20 September 2022.

Ehrsson, H.H. (2007). The Experimental Induction of Out-of-Body Experiences. *Science* 317, 1048.

Elamiryan, R.G. (2019). Human Security in Context of Globalization: Information Security Aspect. www.academia.edu/21857072/human_security_in_context_of_globalization_information_security_aspect. Accessed 11 March 2020.

Eldridge, R., Koser, K., Levin, M., and Rai, S. (2017). What Does the Fourth Industrial Revolution Mean for Migration? https://www.weforum.org/agenda/2017/06/what-does-the-fourth-industrial-revolution-mean-for-migration/. Accessed 17 March 2020.

Fukuda-Parr, S. (2003). New Threats to Human Security in the Era of Globalization. *Journal of Human Development* 4(2), 167–179.

Ghadge, A., Dani, S., and Kalawsky, R. (2012). Supply Chain Risk Management: Present and Future Scope. *The International Journal of Logistics Management* 23, 313–339.

Ghobakhloo, M., Fathi, M., Iranmanesh, M., Maroufkhani, P., and Morales, M.E. (2021). Industry 4.0 Ten Years On: A Bibliometric and Systematic Review of Concepts, Sustainability Value Drivers, and Success Determinants. *Journal of Cleaner Production* 302, 127052.

Gillieron, L. (2019). Davos: Leaders Talk About Globalization as Though It's Inevitable—When It Isn't. https://theconversation.com/davosleaders-talk-about-Globalization-as-though-its-inevitable-when-it-isnt-110216. Accessed 30 February 2020.

Goncharov, V.V. (2020). The Fourth Industrial Revolution: Challenges, Risks and Opportunities. *E-Journal of the World Academy of Art & Science* 2(6), 95–106.

Guo, Y. (2011). Research on Knowledge-Oriented Supply Chain Risk Management System Model. *Journal of Management and Strategy* 2, 72–77.

Gupta, A. (2013). Environment & PEST Analysis: An Approach to External Business Environment. *International Journal of Modern Social Sciences* 2, 34–43.

Habrat, D. (2020). Legal Challenges of Digitalization and Automation in the Context of Industry 4.0. *Procedia Manufacturing* 51, 938–942. https://doi.org/10.1016/j.promfg.2020.10.132.

Ho, W., Zheng, T., Yildiz, H., and Talluri, S. (2015). Supply Chain Risk Management: A Literature Review. *International Journal of Production Research* 53, 5031–5069.

Ignjatović, Đ. (2018). *Kriminologija*. Beograd: Univerzitet u Beogradu, Pravni fakultet.

Kamble, S., Gunasekaran, A., and Dhone, N.C. (2020). Industry 4.0 and Lean Manufacturing Practices for Sustainable Organisational Performance in Indian Manufacturing Companies. *International Journal of Production Research* 58, 1319–1337.

Kodym, O., Kubáč, L., and Kavka, L. (2020). Risks Associated With Logistics 4.0 and Their Minimization Using Blockchain. *Open Engineering* 10, 74–85. https://doi.org/10.1515/eng-2020-0017.

Kuzmenko, G., Skorodumova, O., and Melikov, I. (2018). Basic Needs Determining the Transformation of the System of Education in the Information Age. *Economic and Social Development*, 472–479.

Ljajic, S., Meta, M., and Mladenović, Ž. (2016). Globalizacija: ekonomski i psihološki aspekti. *Ekonomski signali: poslovni magazin* 11(1), 39–62.

Luthra, S., Kumar, A., Zavadskas, E.K., Mangla, S.K., and Garza-Reyes, J.A. (2020). Industry 4.0 as an Enabler of Sustainability Diffusion in Supply Chain: An Analysis of Influential Strength of Drivers in an Emerging Economy. *International Journal of Production Research* 58, 1505–1521.

Magid, E., Zakiev, A., Tsoy, T., Lavrenov, R., and Rizvanov, A. (2021). Automating Pandemic Mitigation. *Advanced Robotics* 35, 572–589.

Mehdi, A. (2013). Dimension of Globalization. https://shodhganga.inflibnet.ac.in/bitstream/10603/24318/9/09_chapter_3.pdf. Accessed 29 March 2020.

Mitrovic, L.M. (2020). Challenges, Risks and Threats to Human Security in the 4th Industrial Revolution. *NBP Journal of Criminalistics and Law* 25(1), 81–97. https://doi.org/10.5937/nabepo25-26316.

Mohebbi, S., Zhang, Q., Wells, E.C., Zhao, T., Nguyen, H., Li, M., Abdel-Mottaleb, N., Uddin, S., Lu, Q., Wakhungu, M.J., and Wu, Z. (2020). Cyber-Physical-Social Interdependencies and Organizational Resilience: A Review of Water, Transportation, and Cyber Infrastructure Systems and Processes. *Sustainable Cities and Society* 62, 102327.

National Institute of Standard and Technology (NIST). (2018). *Framework for Improving Critical Infrastructure Cybersecurity* (pp. 1–55). Gaithersburg, MD: National Institute of Standard and Technology, United States. https://nvlpubs.nist.gov/nistpubs/cswp/nist.cswp.04162018.pdf.

Skorodumova, O.B., and Melikov, I.M. (2020). Social Risks and Cultural Transformation in the Era of Fourth Industrial Revolution. *Conference: International Scientific Conference Social and Cultural Transformations in the Context of Modern Globalism,* dedicated to the 80th Anniversary of Turkayev Hassan Vakhitovich. https://www.researchgate.net/deref/http%3A%2F%2Fdx.doi.org%2F?_tp=eyJjb250ZXh0Ijp7ImZpcnN0UGFnZSI6InB1YmxpY2F0aW9uIiwicGFnZSI6InB1YmxpY2F0aW9uIn19.

Sobb, T., Turnbull, B., and Moustafa, N. (2020). Supply Chain 4.0: A Survey of Cyber Security Challenges, Solutions and Future Directions. *Electronics* 9, 1864.

Sodhi, M.S., Son, B.-G., and Tang, C.S. (2012). Researchers' Perspectives on Supply Chain Risk Management. *Production and Operations Management* 21, 1–13.

Tang, O., and Nurmaya Musa, S. (2011). Identifying Risk Issues and Research Advancements in Supply Chain Risk Management. *International Journal of Production Economics* 133, 25–34.

Williamson, J. (1998). Globalization: The Concept, Causes, and Consequences. www.piie.com/commentary/speeches-papers/Globalization-concept-causes-and-consequences. Accessed 27 February 2020.

World Economic Forum (WEF). (2019). Globalization 4.0: Shaping a New Global Architecture in the Age of the Fourth Industrial Revolution. http://www3.weforum.org/docs/WEF_Globalization_4.0_Call_for_Engagement.pdf. Accessed 17 March 2020.

Zinchenko, Y.P., Menshikov, G.Y., Bayakovsky, Y.M., Chernorizov, A.M., and Voiskunsky, A.E. (2010). Virtual Reality Technologies: Methodological Aspects, Achievements, and Prospects. *National Psychological Journal* 1(3), 54–62.

Index

For Product Safety Concerns and Information please contact our EU
representative GPSR@taylorandfrancis.com
Taylor & Francis Verlag GmbH, Kaufingerstraße 24, 80331 München, Germany

www.ingramcontent.com/pod-product-compliance
Ingram Content Group UK Ltd.
Pitfield, Milton Keynes, MK11 3LW, UK
UKHW021825240425
457818UK00006B/75

* 9 7 8 1 0 3 2 7 1 3 7 7 9 *